Ernest Ingersoll

Golden Alaska

A complete account to date of the Yukon Valley; its history, geography, mineral and

other resources, opportunities and means of access. Vol. 1

Ernest Ingersoll

Golden Alaska
A complete account to date of the Yukon Valley; its history, geography, mineral and other resources, opportunities and means of access. Vol. 1

ISBN/EAN: 9783337310691

Printed in Europe, USA, Canada, Australia, Japan

Cover: Foto ©ninafisch / pixelio.de

More available books at **www.hansebooks.com**

GOLDEN ALASKA

A COMPLETE ACCOUNT TO DATE

OF THE

YUKON VALLEY

*ITS HISTORY, GEOGRAPHY, MINERAL AND OTHER
RESOURCES, OPPORTUNITIES AND
MEANS OF ACCESS*

ERNEST INGERSOLL,
(Formerly with the Hayden Survey in the West)

AUTHOR OF

"KNOCKING 'ROUND THE ROCKIES" "THE CREST OF THE CONTINENT,"
ETC., AND GENERAL EDITOR OF RAND, McNALLY &
Co.'s "GUIDE BOOKS."

CHICAGO AND NEW YORK:
RAND, McNALLY & COMPANY.
1897.

INTRODUCTION.

To make "a book about the Klondike" so shortly after that word first burst upon the ears of a surprised world, would be the height of literary impudence, considering how remote and incommunicado that region is, were it not the public is intensly curious to know whatever can be said authentically in regard to it. "The Klondike," it must be remembered, is, in reality, a very limited district—only one small river valley in a gold-bearing territory twice as large as New England; and it came into prominence so recently that there is really little to tell in respect to it because nothing has had time to happen and be communicated to the outside world. But in its neighborhood, and far north and south of it, are other auriferous rivers, creeks and bars, and mountains filled with untried quartz-ledges, in respect to which information has been accumulating for some years, and where at any moment "strikes" may be made that shall equal or eclipse the wealth of the Klondike placers. It is possible, then, to give here much valuable information in

regard to the Yukon District generally, and this the
writer has attempted to do. The best authority for
early exploration and geography is the monumen-
tal work of Capt. W. H. Dall, "Alaska and its Resources," whose companion, Frederick Whymper,
also wrote a narrative of their adventures. The
reports of the United States Coast Survey in that
region, of the exploration of the Upper Yukon by
Schwatka and Hayes of the United States Geological
Survey, of Nelson, Turner and others attached to
the Weather Service, of the Governor of the Territory, of Raymond, Abercrombie, Allen and other
army and naval officers who have explored the
coast country and reported to various departments
of the government, and of several individual explorers, especially the late E. J. Glave, also contain facts
of importance for the present compilation. The
most satisfactory sources of information as to the
geography, routes of travel, geology and mineral-
ogy and mining development, are contained in the
investigations conducted some ten years ago by the
Canadian Geological Survey, under the leadership
of Dr. G. M. Dawson and of William Ogilvie. Of
these I have made free use, and wish to make an
equally free acknowledgement.

It will thus be found that the contents of this
pamphlet justified even the hasty publication which

the public demands, and which precludes much attention to literary form; but an additional claim to
attention is the information it seeks to give intend-
ing travelers to to that far-away and very new and as
yet unfurnished region, how to go and what to
take, and what are the conditions and emergencies
which they must prepare to meet. Undoubtedly the
pioneers to the Yukon pictured the difficulties of
the route and the hardships of their life in the highest colors, both to add to their self-glory and to re-
duce competition. Moreover, every day mitigates
the hardships and makes easier the travel. Nevertheless, enough difficulties, dangers and chances of
failure remain to make the going to Alaska a matter
for very careful forethought on the part of every
man. To help him weigh the odds and choose
wisely, is the purpose of this little book.

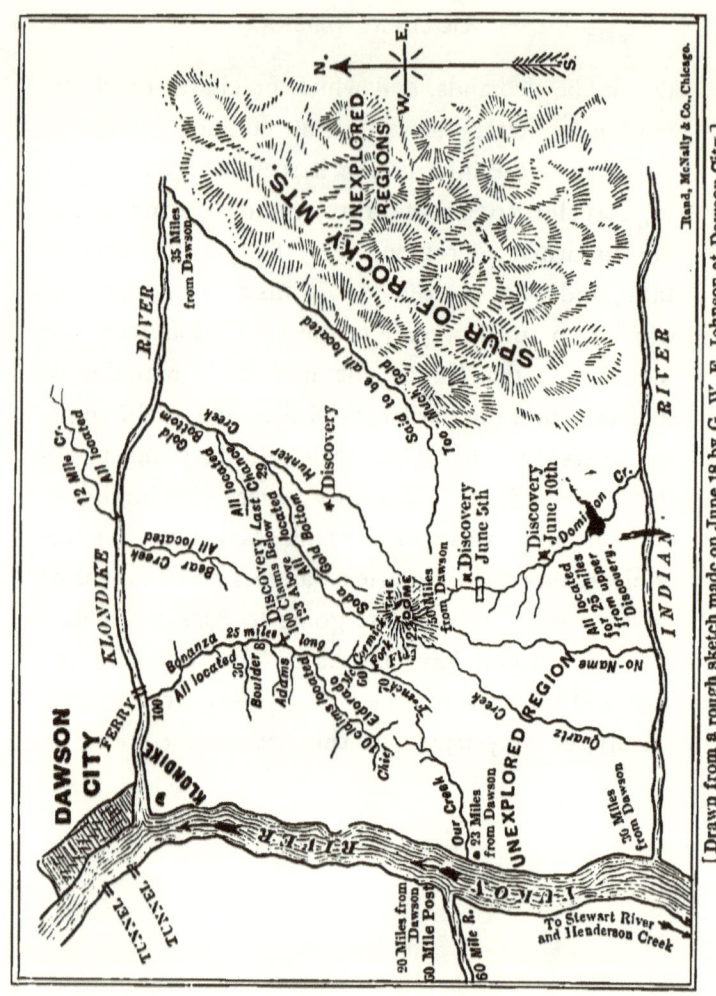

[Drawn from a rough sketch made on June 18 by G. W. F. Johnson at Dawson City.]

Rand, McNally & Co., Chicago.

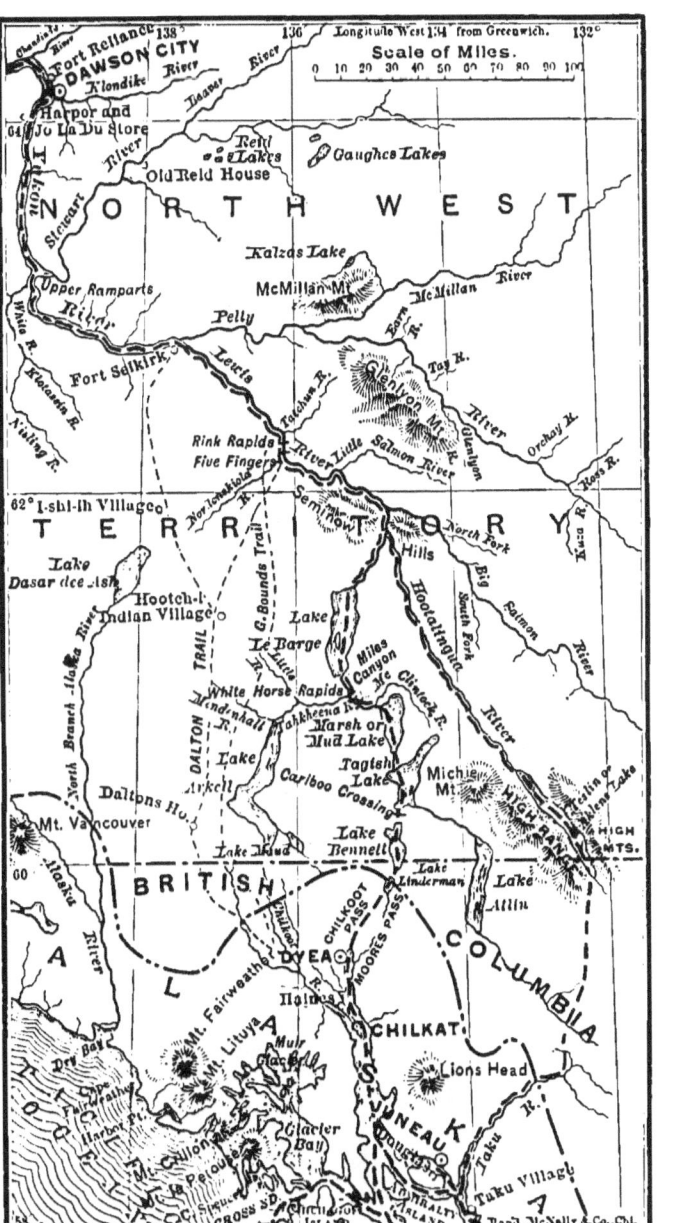

BIRD'S-EYE VIEW OF SITKA—FROM BARANOFF CASTLE.

GOLDEN ALASKA.

ROUTES TO THE YUKON GOLD-FIELDS.

The gold-fields of the Yukon Valley, at and near
Klondike River, are near the eastern boundary of
Alaska, from twelve to fifteen hundred miles up from
the mouth of the river, and from five to eight hundred miles inland by the route across the country
from the southern Alaskan coast. In each case an
ocean voyage must be taken as the first step; and
steamers may be taken from San Francisco, Portland, Ore., Seattle, Wash., or from Victoria, B. C.

The overland routes to these cities require a
word.

1. To San Francisco. This city is reached di-
rectly by half a dozen routes across the plains and
Rocky Mountains, of which the Southern Pacific,
by way of New Orleans and El Paso; the Atchison
& Santa Fé and Atlantic & Pacific by way of
Kansas City, and across northern New Mexico and
Arizona; the Burlington, Denver & Rio Grande, by
way of Denver and Salt Lake City; and the Union

and Central Pacific, by way of Omaha, Ogden and
Sacramento, are the principal ones.

2. To Portland, Oregon. This is reached directly
by the Union Pacific and Oregon Short Line, via
Omaha and Ogden; and by the Northern Pacific,
via St. Paul and Helena, Montana.

3. To Seattle, Wash. This city, Tacoma, Port
Townsend and other ports on Puget Sound, are
the termini of the Northern Pacific Railroad and
also of the Great Northern Railroad from St. Paul
along the northern boundary of the United States.
The Canadian Pacific will also take passengers there
expeditiously by rail or boat from Vancouver, B. C.

4. To Vancouver and Victoria, B. C. Any of
the routes heretofore mentioned reach Victoria by
adding a steamboat journey; but the direct route,
and one of the pleasantest of all the transcontinental
routes, is by the Canadian Pacific Railway from
Montreal or Chicago, via Winnipeg, Manitoba, to
the coast at Vancouver, whence a ferry crosses to
Victoria.

Regular routes of transportation to Alaska are
supplied by the Pacific Coast Steamship Company,
which has been dispatching mail-steamships once a
fortnight the year round from Tacoma to Sitka,
which touch at Juneau and all other ports of call.
They also maintain a service of steamers between

San Francisco and Portland and Puget Sound ports. These are fitted with every accommodation and luxury for tourist-travel; and an extra steamer, the Queen, has been making semi-monthly trips during June, July and August. These steamers would carry 250 passengers comfortably and the tourist fare for the round trip has been $100.

The Canadian Pacific Navigation Company has been sending semi-monthly steamers direct from Victoria to Port Simpson and way stations the year round. They are fine boats, but smaller than the others and are permitted to land only at Sitka and Dyea.

Such are the means of regular communication with Alaskan ports. There has been no public conveyance north of Sitka, except twice or thrice a year in summer in the supply-steamers of the Alaskan commercial companies, which sailed from San Francisco to St. Michael and there transferred to small boats up the Yukon.

Whether any changes will be made in these schedules for the season of 1898 remains to be seen.

Special steamers.—As the regular accommodations were found totally inadequate to the demand for passage to Alaska which immediately followed the report of rich discoveries on Klondike Creek,

extra steamers were hastily provided by the old
companies, others are fitted up and sent out by
speculative owners, and some have been privately
chartered. A score or more steamships, loaded with
passengers, horses, mules and burros (donkeys) to
an uncomfortable degree, were thus despatched from
San Francisco, Puget Sound and Victoria between
the middle of July and the middle of August. An
example of the way the feverish demand for transportation is found in the case of the Willamette,
a collier, which was cleaned out in a few hours and
turned into an extemporized passenger-boat. The
whole 'tween decks space was filled with rough
bunks, wonderfully close together, for "first-class"
passengers; while away down in the hold second-
class arrangements were made which the mind shud-
ders to contemplate. Yet this slave-ship sort of a
chance was eagerly taken, and such space as was
left was crowded with animals and goods. Many
persons and parties bought or chartered private
steamers, until the supply of these was exhausted
by the end of August.

Two routes may be chosen to the gold-fields.

1. By way of the Yukon River. This is all the
way by water, and means nearly 4,500 miles of voy-
aging.

2. By way of the seaports of Dyea or Shkagway,

GLACIER BAY.

STEAMSHIP QUEEN.

over mountain passes, afoot or a-horseback, and down the upper Yukon River and down the lakes and rivers by raft, skiff and steamboat.

To describe these routes is the next task—first, that by the way of St. Michael, and second—up the Yukon River.

Route, via St. Michael and the Yukon River.—This begins by a sea-voyage, which may be direct, or along the coast. The special steamers (and future voyages, no doubt) usually take a direct course across the North Pacific and through the Aleutian Islands to St. Michael, in Norton Sound, a bight of Bering Sea. The distance from San Francisco is given as 2,850 miles; from Victoria or Seattle, about 2,200 miles. The inside course would be somewhat longer, would follow the route next to be described as far as Juneau and Sitka, then strike northwest along the coast to St. Michael.

This town, on an island near shore in Norton Sound, was established in 1835 by Lieut. Michael Tébenkoff, of the Russian navy, who named it after his patron saint. Though some distance to the mouth of the Yukon entrance, St. Michael has always been the controlling center and base of supplies for the great valley. The North American Trading and Transportation Company and the Alaska Commercial Company have their large warehouses here, and provide the miners with tools, clothing and provisions. Recently the wharf and warehouse accommodations have been extended, and the population has increased, but if, as is probable, any considerable number of men are stopped there this fall by the freezing of the river, and compelled to pass the winter on the island, they will find it a dreary, if not dangerous experience.

The vessels supplying this depot can seldom approach the anchorage of St. Michael before the end of June on account of large bodies of drifting ice that beset the waters of Norton Sound and the straits between St. Lawrence and the Yukon Delta.

A temporary landing-place is built out into water deep enough for loaded boats drawing five feet to come up at high tide, this is removed when winter approaches, as otherwise it would be destroyed by ice. The shore is sandy and affords a moderately sloping beach, on which boats may be drawn up. A few feet only from high water mark are perpendicular banks from six to ten feet high, composed of decayed pumice and ashes, covered with a layer about four feet thick of clay and vegetable matter resembling peat. This forms a nearly even meadow with numerous pools of water, which gradually ascends for a mile or so to a low hill, of volcanic origin, known as the Shaman Mountain.

Between the point on which St. Michael is built and the mainland, a small arm of the sea makes in, in which three fathoms may be carried until the flagstaff of the fort bears west by north, this is the best-protected anchorage, and has as much water and as good bottom as can be found much farther out.

The excitement of the summer of 1897 caused an enlargement of facilities and the erection of additional buildings, forming a nucleus of traffic called Fort Get There. Here will be put together in the autumn or winter at least three, and perhaps more, new river steamboats, of which only two or three have been running on the lower river during the last two or three years. These are taken up, in pieces, by ships and fitted together at this point. All are flat-bottomed, stern-wheeled, powerfully engined craft, the largest able to carry perhaps 250 tons, such as run on the upper Missouri, and they will burn wood, the cutting and stacking of which on the river bank will furnish work to many men during the coming winter. To such steamers, or smaller boats, all the persons and cargoes must be transferred at St. Michael.

For the last few years there has been no trader here but the agent of the Alaska Commercial Company, and a story is told of the building of a river-

boat there in 1892, which illustrates what life on
the Yukon used to be.   In that year a Chicago man,
P. B. Weare, resolved to enter the Alaskan field as
a trader.   He chartered a schooner, and placed upon
it a steamboat, built in sections and needing only to
be put together and have its machinery set up, and
for·this purpose he took with him a force of carpenters and machinists.   On reaching St. Michael
Weare was refused permission to land his boat sections on the land of the Commercial Company's
post, and was compelled to make a troublesome
landing on the open beach, where he began operations.   Suddenly his ship carpenters stopped work.
They had been offered, it was said, double pay by
the rival concern if they would desist from all work.
Weare turned to the Indians, but with the same ill-
success.   The Indians were looking out for their
winter grub.   Here was the Chicago man 2,500
miles from San Francisco and only two weeks
left to him in which to put his boat together and
then hope for a chance to ascend the river before
winter came on.   There was no time in which to
get additional men from San Francisco.   In the
midst of his trouble Weare one day espied the rev-
enue cutter Bear steaming into the roadstead.   On
board of her was Captain Michael A. Healy.   That
officer, on going ashore and discovering the con-

STEAMER PORTUS B. WEARE.

dition of affairs, threatened to hang every carpenter
and mechanic Weare had brought up if they failed
to immediately commence work. The men went to
work, and with them went a gang of men from the
Bear. The little steamer was put together in a few
days, and the Bear only went to sea after seeing
the P. B. Weare steaming into the mouth of the
Yukon.

The Weare was enabled that summer to land her
stores along the Yukon, and was the only vessel
available for the early crowds of miners going to
Klondike.

The mouth of the Yukon is a great delta, surrounded by marsh of timber—a soaking prairie in
summer, a plain of snow and ice in winter. The
shifting bars and shallows face out from this delta
far into Bering Sea, and no channel has yet been
discovered whereby an ocean steamer could enter
any of the mouths. Fortunately the northernmost
mouth, nearest St. Michael and 65 miles from it, is
navigable for the light river steamers, and this one,
called Aphoon, and marked by its unusual growth
of willows and bushes is well known to the local
Russian and Indian pilots. It is narrow and intricate, and the general course up stream is south-
southeast. Streams and passages enter it, and it has
troublesome tidal currents. The whole space be-

tween the mouth is a net-work, indeed, of narrow channels, through the marshes.

Kutluck, at the outlet of the Aphoon, on Pastol Bay, is an Indian village ,long celebrated for its manufacture of skin boats (bidars), and there the old-time voyagers were accustomed to get the only night's sleep ashore that navigation permits between St. Michael and Andraefski. On the south bank of the main stream, at the head of the delta, is the Roman Catholic mission of Kuslivuk; and a few miles higher, just above the mouth of the Andraefski River, is tne abandoned Russian trading post, Andraefski, above which the river winds past Icogmute, where there is a Greek Catholic mission. The banks of the river are much wooded, and the current even as far down as Koscrefski averages over three knots an hour. Above Koserefski (the Catholic Mission station), the course is along stretches of uninviting country, among marsh islands and "sloughs," the current growing more and more swift on the long reach from Auvik, where the Episcopal mission is situated, to Nulato.

The river here has a nearly north and south course, parallel with the coast of Norton Sound and within fifty miles or so of it. Two portages across here form cut-offs in constant use in winter by the traders, Indians and missionaries. The first of these portages starts from the mainland opposite the Island of St. Michael, and passes over the range of hills that defines the shore to the headwaters of the Anvik River. This journey may be made in winter by sledges and thence down the Auvik to the Yukon, but it is a hard road. Mr. Nelson, the naturalist, and a fur trader, spent two months from November 16, 1880, to January 19, 1891, in reaching the Yukon by this path.

The other portage is that between Unalaklik, a Swedish mission station at the mouth of the Unalaklik River, some fifty miles north of St. Michael, and a stream that enters the Yukon half way between Auvik and Nulato. In going from St. Michael to Unalatlik there are few points at which a boat can land even in the smoothest weather; in rough weather only Major's Cove and Kegiktowenk before rounding Tolstoi Point to Topánika, where there is a trading post. Topánika is some ten miles from Unalaklik, with a high shelving beach, behind which rise high walls of sandstone in perpendicular bluffs from twenty to one hundred feet in height. This beach continues all the way to the Unalaklik River, the bluff gradually decreasing into a marshy plain at the river's mouth, which is obstructed by a bar over which at low tide there are only a few feet of water except in a narrow and tortuous channel, constantly changing as the river deposits fresh detritus. Inside this bar there are two or three fathoms for a few miles, but the channel has only a few feet, most of the summer, from the mouth of the river to Ulu-kuk.

Trees commence along the Unalaklik River as soon as the distance from the coast winds and salt air permit them to grow; willow, poplar, birch and spruce being those most frequently found.

The Unalaklik River is followed upward to Ulu-kuk, where begins a sledging portage over the marshes to the Ulukuk Hills, where there is a native village known as Vesolia Sopka, or Cheerful Peak, at an altitude of eight hundred feet above the surrounding plain. This is a well-known trapping ground, the fox and marten being very plentiful. From Sopka Vesolia (Cheerful Peak) it is about one day's journey to Beaver Lake, which is only a marshy tundra in winter, but is flooded in the spring and summer months. From the high hills beyond the lake one may catch a first glimpse of the great Yukon sweeping between its splendid banks.

The natives call Nulato emphatically a "hungry" place, and it was once the scene of an atrocious massacre. Capt. Dall, from whose book much of the information regarding this part of Alaska is derived, describes the Indians here as a very great nuisance.

OLD RUSSIAN BLOCK HOUSE AT SITKA.

"They had," he explains, "a great habit of coming in and sitting down, doing and saying nothing, but watching everything. At meal times they seemed to count and weigh every morsel we ate, and were never backward in assisting to dispose of the remains of the meal. Occasionally we would get desperate and clean them all out, but they would drop in again and we could do nothing but resign ourselves."

The soil on the banks of the Yukon and that of the islands probably never thaws far below the surface. It is certain that no living roots are found at a greater depth than three feet. The soil, in layers that seems to mark annual inundations, consists of a stratum of sand overlaid by mud and covered with vegetable matter, the layers being from a half inch to three inches in thickness. In many places where the bank has been undermined these layers may be counted by the hundred. Low bluffs of blue sandstone, with here and there a high gravel bank, characterize the shores as far as Point Sakataloutan, and some distance above this point begin the quartzose rocks.

The next station on the river is the village of Nowikakat, on the left bank. Here may be obtained stores of dried meat and fat from the Indians. The village is situated upon a beautiful bay or Nowika-

kat Harbor, which is connected by a narrow entrance with the Yukon. "Through this a beautiful
view is obtained across the river, through the numerous islands of the opposite shore, and of the Yukon
Mountains in the distance. The feathery willows
and light poplars bend over and are reflected in the
dark water, unmixed as yet with Yukon mud;
every island and hillside is clothed in the delicate
green of spring, and luxuriates in a density of foli-
age remarkable in such a latitude."

Nowikakat is specially noted for the excellence of
its canoes, of which the harbor is so full that a boat
makes its landing with difficulty among them. It is
the only safe place on the lower Yukon for winter-
ing a steamer, as it is sheltered from the freshets
which bring down great crushes of ice in the spring.

At Nuklukahyet there is a mission of the Episcopal church and a trading store, but there may or
may not be supplies of civilized goods, not to speak
of moose meat and fat. This is the neutral ground
where all the tribes meet in the spring to trade.
The Tananah, which flows into the Yukon at this
point, is much broader here than the Yukon, and
it is here that Captain Dall exclaims in his diary:
"And yet into this noble river no white man has
dipped his paddle." Recently, however, the Tananah has been more or less explored by prospectors

with favorable results towards the head of the river, which is more easily reached overland from Circle City and the Birch Creek camps.

Leaving Nuklukahyet, the "Ramparts" are soon sighted, and the Yukon rapids sweep between bluffs and hills which rise about fifteen hundred feet above the river, which is not more than half a mile wide and seems almost as much underground as a river bed in a canyon. The rocks are metaphoric quartzites, and the river-bed is crossed by a belt of granite. The rapid current has worn the granite away at either side, making two good channels, but in the center lies an island of granite over which the water plunges at high water, the fall being about twelve feet in half a mile.

Beyond the mouth of the Tananah the Yukon begins to widen, and it is filled with small islands. The mountains disappear, and just beyond them the Totokakat, or Dall River of Ketchum, enters the Yukon from the north. Beyond this point the river, ever broadening, passes the "Small Houses," deserted along the bank at the time, years ago, when the scarlet fever, brought by a trading vessel to the mouth of the Chilkat, spread to the Upper Yukon and depopulated the station. This place is noted for the abundance of its game and fish.

The banks of the river above this point become

very low and flat, the plain stretching almost un-
broken to the Arctic Ocean.

The next stream which empties into the Yukon
is Beaver Creek, and farther on the prospector
bound for Circle City may make his way some two
hundred miles up Birch Creek, along which much
gold has already been discovered, to a portage of
six miles, which will carry him within six miles of
Circle City on the west.

Meanwhile the Yukon passes Porcupine River and
Fort Yukon, the old trading-post founded in 1846-7,
about a mile farther up the river than the present
fort is situated. The situation was changed in 1864,
owing to the undermining of the Yukon, which
yearly washed away a portion of the steep bank until the foundation timbers of the old Redoubt over-
hung the flood.

Many small islands encumber the river from Fort
Yukon to Circle City, and the river flows along the
rich lowland to the towns and mining centers of the
new El Dorado, an account of which belongs to
a future chapter.

This voyage can be made only between the middle
of June and the middle of September, and requires
about forty days, at best, from San Francisco to Circle City or Forty Mile.

Route via Juneau, the Passes and down the Up-

INDIAN TOTEM POLE, FORT SIMPSON

per Yukon River. The second and more usual, because shorter and quicker course, is that to the
head of Lynn Canal (Taiya Inlet) and overland.
This coast voyage may be said to begin at Victoria,
B. C. (since all coast steamers gather and stop
there), where a large number of persons prefer to
buy their outfits, since by so doing, and obtaining
a certificate of the fact, they avoid the custom duties exacted at the boundary line on all goods and
equipments brought from the United States. Victoria is well supplied with stores, and is, besides, one
of the most interesting towns on the Pacific coast.
The loveliest place in the whole neighborhood is
Beacon Hill Park, and is well worth a visit by those
who find an hour or two on their hands before the
departure of the steamer. It forms a half-natural,
half-cultivated area of the shore of the Straits of
Fuca, where coppices of the beautiful live oak, and
many strange trees and shrubs mingled with the all-
pervading evergreens.

Within three miles of the city, and reached by
street cars, is the principal station in the North Pacific of the British navy, at Esquimault Bay. This
is one of the most picturesque harbors in the world,
and a beginning is made of fortifications upon a very
large scale and of the most modern character. This
station, in many respects, is the most interest-

ing place on the Pacific coast of Canada.

Leaving Victoria, the steamer makes its way cautiously through the sinuous channels of the harbor into the waters of Fuca Strait, but this is soon left behind and the steamer turns this way, and that, at the entrance to the Gulf of Georgia, among those islands through which runs the international boundary line, and for the possession of which England and the United States nearly went to war in 1862. The water at first is pale and somewhat opaque, for it is the current of the great Fraser gliding far out upon the surface, and the steamer passes on beyond it into the darker, clearer, salter waters of the gulf. Then the prow is headed to Vancouver, where the mails, freight and new railway passengers are received.

From Vancouver the steamer crosses to Nanaimo, a large settlement on Vancouver Island, where coal mines of great importance exist. A railway now connects this point with Victoria, and a wagon road crosses the interior of the island to Alberni Canal and the seaport at its entrance on Barclay Sound. This is the farthest northern telegraph point. The mines at Nanaimo were exhausted some time ago, after which deep excavations were made on Newcastle Island, just opposite the town. But after a tremendous fire these also were abandoned, and all

the workings are now on the shores of Departure
Bay, where a colliery village named Wellington has
been built up. A steam ferry connects Nanaimo
with Wellington; and while the steamer takes in its
coal, the passengers disperse in one or the other village, go trout fishing, shooting or botanizing in the
neighboring woods, or trade and chaffer with the
Indians. Nanaimo has anything but the appearance of a mining town. The houses do not stretch
out in the squalid, soot-covered rows familiar to
Pennsylvania, but are scattered picturesquely, and
surrounded by gardens.

Just ahead lie the splendid hills of Texada Island,
whose iron mines yield ore of extraordinary purity,
which is largely shipped to the United States to be
made into steel. The steamer keeps to the left, making its way through Bayne's Sound, passing Cape
Lazaro on the left and the upper end of Texada on
the right, across the broadening water along the
Vancouver shore into Seymour Narrows. These
narrows are only about 900 yards wide, and in them
there is an incessant turmoil and bubbling of currents. This is caused by the collision of the streams
which takes place here; the flood stream from the
south, through the Strait of Fuca and up the Haro
Archipelago being met by that from Queen Charlotte Sound and Johnstone straits. These straits are

about 140 miles long, and by the time their full
length is passed, and the maze of small islands
on the right and Vancouver's bulwark on the left
are escaped together, the open Pacific shows itself
for an hour or two in the offing of Queen Charlotte's
Sound, and the steamer rises and falls gently upon
long, lazy rollers that have swept all the way from
China and Polynesia. Otherwise the whole voyage
is in sheltered waters, and seasickness is impossible.
The steamer's course now hugs the shore, turning
into Fitz Hugh Sound, among Calvert, Hunter's
and Bardswell islands, where the ship's spars sometimes brush the overhanging trees. Here are the
entrances to Burke Channel and Dean's Canal that
penetrate far amid the tremendous cliffs of the mainland mountains. Beyond these the steamer dashes
across the open bight of Milbank Sound only to enter the long passages behind Princess Royal, Pit
and Packer islands, and coming out at last into
Dixon Sound at the extremity of British Columbia's
ragged coast line.

The fogs which prevail here are due to the fact
that this bight is filled with the waters of the warm
Japanese current and the gulf stream of the Pacific
from which the warm moisture rises to be condensed
by the cool air that descends from the neighboring
mountains, into the dense fogs and heavy rain

STREET IN SITKA.

storms to which the littoral forest owes its extraordinary luxuriance. During the mid-summer and
early autumn, however, the temperature of air and
water become so nearly equable that fog and rain
are the exception rather than the rule.

Crossing the invisible boundary into Alaska the
steamer heads straight toward Fort Tougass, on
Wales Island, once a military station of the United
States, but now only a fishing place. Between this
point and Fort Wrangel another abandoned military
post of the United States, two or three fish canneries
and trading stations are visited and the ship goes on
among innumerable islands and along wide reaches
of sound to Taku Inlet (which deeply indents the
coast, and is likely in the near future to become an
important route to the gold fields), and a few hours
later Juneau City is reached.

Juneau City has been lately called the key to the
Klondike regions, as it is the point of departure for
the numberless gold hunters who, when the season
opens again, will rush blindly over incalculably rich
ledges near the coast to that remote inland El Dorado of their dreams.

Juneau has for seventeen years been supported by
the gold mines of the neighboring coast. It is situated ten miles above the entrance of Gastineau Channel, and lies at the base of precipitous mountains,

its court house, hotels, churches, schools, hospital and opera house forming the nucleus for a population which in 1893 aggregated 1,500, a number very largely increased each winter by the miners who gather in from distant camps. The saloons, of which in 1871 there were already twenty-two, have increased proportionately, and there are, further, at least one weekly newspaper, one volunteer fire brigade, a militia company and a brass band in Juneau. The curio shops on Front and Seward streets are well worth visiting, and from the top of Seward Street a path leads up to the Auk village, whose people claim the flats at the mouth of Gold Creek. A curious cemetery may be seen on the high ground across the creek, ornamented with totemic carvings and hung with offerings to departed spirits which no white man dares disturb.

FROM JUNEAU TO THE GOLD FIELDS.

The few persons who formerly wished to go to the head of Lynn Canal did so mainly by canoeing, or chartered launches, but now many opportunities are offered by large steamboats. Most of the steamers that bring miners and prospectors from below do not now discharge their freight at Juneau, however, but go straight to the new port Dyea at the head of the canal. Lynn Canal is the grandest fiord on the coast, which it penetrates for seventy-five

miles. It is then divided by a long peninsula called
Seduction Point, into two prongs, the western of
which is called Chilkat Inlet, and the eastern Chilkoot. "It has but few indentations, and the abrupt
palisades of the mainland shores present an unri-
valled panorama of mountains, glaciers and forests,
with wonderful cloud effects. Depths of 430 fathoms have been sounded in the canal, and the continental range on the east and the White Mountains
on the west rise to average heights of 6,000 feet,
with glaciers in every ravine and alcove." No Cam-
eron boundary line, which Canada would like to es-
tablish, would cut this fiord in two, and make it use-
less to both countries in case of quarrel. The magnificent fan-shaped Davidson glacier, here, is only
one among hundreds of grand ice rivers shedding
their bergs into its waters. At various points salmon canneries have long been in operation; and the
Seward City mines are only the best among several
mineral locations of promise. A glance at the map
will show that this "canal" forms a straight continu-
ation of Chatham Strait, making a north and south
passage nearly four hundred miles in length, which
is undoubtedly the trough of a departed glacier.

Dyea, the new steamer landing and sub-port of
entry, is at the head of navigation on the Chilkoot
or eastern branch of this Lynn Canal, and takes its

name, in bad modern spelling, from the long-known Taiya Inlet, which is a prolongation inland for twenty miles of the head of the Chilkoot Inlet. It should continue to be spelled Tiaya. This inlet is far the better of the two for shipping, Chilkat Inlet being exposed to the prevalent and often dangerous south wind, so that it is regarded by navigators as one of the most dangerous points on the Alaskan coast. A Presbyterian mission and government school were formerly sustained at Haines, on Seduction Point, but were abandoned some years ago on account of Indian hostility.

The Passes.—Three passes over the mountains are reached from these two inlets,—Chilkat, Chilkoot and White.

Chilkat Pass is that longest known and formerly most in vogue. The Chilkat Indians had several fixed villages near the head of the inlet, and were accustomed to go back and forth over the mountains to trade with the interior Indians, whom they would not allow to come to the coast. They thus enjoyed not only the monopoly of the business of carrying supplies over to the Yukon trading posts and bringing out the furs, and more recently of assisting the miners, but made huge profits as middlemen between the Indians of the interior and the trading posts on the coast. They are a sturdy race

HEAD WATERS, DYEA RIVER.

of mountaineers, and the most arrogant, treacherous and turbulent of all the northwestern tribes, but their day is nearly passed. The early explorers—Krause, Everette and others—took this pass, and it was here that E. J. Glave first tried (in 1891) to take pack horses across the mountains, and succeeded so well as to show the feasibility of that method of carriage, which put a check upon the extortion and faithlessness of the Indian carriers. His account of his adventures in making this experiment, over bogs, wild rocky heights, snow fields, swift rivers and forest barriers, has been detailed in The Century Magazine for 1892, and should be read by all interested. "No matter how important your mission," Mr. Glave wrote, "your Indian carriers, though they have duly contracted to accompany you, will delay your departure till it suits their convenience, and any exhibition of impatience on your part will only remind them of your utter dependency on them; and then intrigue for increase of pay will at once begin. While en route they will prolong the journey by camping on the trail for two or three weeks, tempted by good hunting or fishing. In a land where the open season is so short, and the ways are so long, such delay is a tremendous drawback. Often the Indians will carry their loads some part of the way agreed on, then demand an extravagant increase of pay or a goodly share of the white man's stores, and, failing to get either, will fling down their packs and return to their village, leaving their white employer helplessly stranded.

The usual charge for Indian carriers is $2 a day and board, and they demand the best fare and a great deal of it, so that the white man finds his precious stores largely wasted before reaching his destination. These facts are mentioned, not because it is now necessary to endure this extortion and expense, but to show how little dependence can be placed upon the hope of securing the aid of Indian packers in carrying the goods of prospectors or explorers elsewhere in the interior, and the great expense involved. This pass descends to a series of connected lakes leading down to Lake Labarge and thence by another stream to the Lewes; and it requires twelve days of pack-carrying—far more than is necessary on the other passes. As a consequence, this pass is now rarely used except by Indians going to the Aksekh river and the coast ranges northward.

Chilkoot, Taiya or Parrier Pass.—This is the pass that has been used since 1885 by the miners and others on the upper Yukon, and is still a route of travel. It starts from the head of canoe navigation on Taiya inlet, and follows up a stream valley, gradually leading to the divide, which is only 3,500 feet above the sea. The first day's march is to the foot of the ascent, and over a terrible trail, through heavy woods and along a steep, rocky and often boggy hillside, broken by several deep gullies. The ascent is then very abrupt and over huge masses of fallen rock or steep slippery surfaces of rock in place. At the actual summit, which for seven or eight miles is bare of trees or bushes, the trail leads through a narrow rocky gap, and the whole scene is one of the most complete desolation. Naked granite rocks, rising steeply to partly snow-clad mountains on either side. Descending the inland or north slope is equally bad traveling, largely over wide areas of shattered rocks where the trail may easily be lost. The further valley contains several little lakes and leads roughly down to Lake Lindeman. The distance from Taiya is twenty-three and a half miles, and it is usually made in two days. Miners sometimes cross this pass in April, choosing fine weather, and then continue down the lakes on the ice to some point where they can conveniently camp and wait for the opening of navigation on the Yukon; ordinarily it is unsafe to attempt a return in the autumn later than the first of October.

Lake Lindeman is a long narrow piece of water navigable for boats to its foot, where a very bad river passage leads into the larger Lake Bennett, where the navigation of the Yukon really begins.

"The Chilkoot Pass," writes one of its latest travelers, "is difficult, even dangerous, to those not possessed of steady nerves. Toward the summit there is a sheer ascent of 1,000 feet, where a slip would certainly be fatal. At this point a dense mist overtook us, but we reached Lake Lindeman—the first of a series of five lakes—in safety, after a fatiguing tramp of fourteen consecutive hours through half-melted snow. Here we had to build our own boat, first felling the timber for the purpose. The journey down the lakes occupied ten days, four of which were passed in camp on Lake Bennett, during a violent storm, which raised a heavy sea. The rapids followed. One of these latter, the "Grand Canyon," is a mile long, and dashes through walls of rock from 50 to 100 feet high; six miles below are the "White Horse Rapids," a name which many fatal accidents have converted into the "Miner's Grave." But snags and rocks are everywhere a fruitful source of danger on this river, and from this rapid downward scarcely a day passed that one did not see some cairn or wooden cross marking the last resting place of some drowned pilgrim to the land of gold. The above is a brief sketch of the troubles that beset the Alaskan gold prospector—troubles that, although unknown in the eastern states and Canada, have for

RAFT ON LAKE LINDEMAN.

many years past associated the name of Yukon with
an ugly sound in western America."

It is probable that few if any persons need go over
this pass next year, and its hardships will become a
tradition instead of a terrible prospect.

White Pass.—This pass lies south of the Chilkoot, and leaves the coast at the mouth of the
Shagway river, five miles south of Dyea and 100
from Juneau. It was first explored in 1887 and was
found to run parallel to the Chilkoot. The distance
from the coast to the summit is seventeen miles, of
which the first five are in level bottom land, thickly
timbered. The next nine miles are in a cañon-like
valley, beyond which three miles, comparatively
easy, take one to the summit, the altitude of which
is roughly estimated at 2,600 feet. Beyond the summit a wide valley is entered and leads gradually to
the Tahko arm of Tahgish lake. This pass, though
requiring a longer carriage, is lower and easier than
the others, and already a pack-trail has been built
through it which will soon be followed by a wagon
road, and surveys for a narrow guage railway are
in progress. At the mouth of the Shkagway River
ocean steamers can run up at all times to a wharf
which has been constructed in a sheltered position,
and there is an excellent town site with protection
from storms.

An English company, the British Columbia Development Association, Limited, has already established a landing wharf and is erecting a wharf and
sawmills at Skagway, whence it is proposed (as soon
as feasible) to lay down a line of rail some thirty-
five miles long, striking the Yukon River at a branch
of the Marsh Lake, about 100 miles below Lake Lin-
demann. By this means the tedious and difficult
navigation between these two points will be avoided,
and the only dangerous parts of the river below will
be circumvented by a road or rail portage. Light-
draught steamers will be put on from Teslin Lake to
the cañon and from the foot of the latter to all the
towns and camps on the river.

Dyea is a village of cabins and tents, and little if
anything in the way of supplies can be got there;
it is a mere forwarding point.

Pending the completion of the facilities mentioned
above, miners may transport their goods over the
pack trail on their own or hired burros, and at Tahgish Lake take a boat down the Tahco arm (11 miles)
to the main lake, and down that lake and its outlet
into Lake Marsh. This chain of lakes, filling the
troughs of old glacial fiords to a level of 2,150 feet
above the sea, "constitutes a singularly picturesque
region, abounding in striking points of view and in
landscapes pleasing in their variety or grand and im-

pressive in this combination of rugged mountain
forms." All afford still-water navigation, and as soon
as the road through White Pass permits the transportation of machinery, they will doubtless be well
supplied with steamboats. Marsh Lake is 20 miles
long, Bennett 26, and Tagish 16½ miles, with Windy
Arm 11 miles long, Tahko Arm 20 miles, and other
long, narrow extensions among the terraced, ever-
green-wooded hills that border its tranquil surface.
The depression in which this group of lakes lies is
between the coast range and the main range of the
Rockies; and as it is sheltered from the wet sea-
winds by the former heights, its climate is nearly as
dry of that of the interior. The banks are fairly well
timbered, though large open spaces exist, and
abound in herbage, grass and edible berries. Lake
Marsh, named by Schwatka after Prof. O. C. Marsh
of Yale, but called Mud Lake by the miners, without
good reason, is twenty miles long and about two
wide. It is rather shallow and the left bank should
be followed. The surrounding region is rather low,
rising by terraces to high ranges on each side, where
Michie mountain, 5,540 feet in height, eastward, and
Mounts Lorue and Landsdowne, westward, 6,400
and 6,140 feet high respectively, are the most prominent peaks. "The diversified form of the mountains in view from this lake render it particularly

picturesque," remarks Dr. Dawson, "and at the time of our visit, on the 10th and 11th of September, the autumn tints of the aspens and other deciduous trees and shrubs, mingled with the sombre greens of the spruces and pines, added to its beauty."

Near the foot of this lake enters the McClintock river, of which little is known. The oulet is a clear, narrow, quiet stream, called Fifty-mile River, which flows somewhat westerly down the great valley. Large numbers of dead and dying salmon are always seen here in summer, and as these fish never reach Lake Marsh, it is evident that the few who are able, after their long journey, to struggle up the rapids, have not strength left to survive.

The descent of the Lewes (or Yukon) may be said to begin at this point, and 23 miles below Lake Marsh the first and most serious obstacle is encountered in the White Horse Rapids, and Miles Cañon. Their length together is 2¾ miles, and they seem to have been caused by a small local effusion of lava, which was most unfortunately ejected right in the path of the river. The cañon is often not more than 100 feet in width, and although parts of it may be run at favorable times, all of it is dangerous, and the White Horse should never be attempted. The portage path in the upper part of the cañon is on the east bank, and is about five-eighths of a mile

DOG PACK TRAIN.

long. There a stretch of navigation is possible, with
caution, ending at the head of White Horse Rapids,
where one must land on the west bank, which consists of steep rocks, very awkward for managing a
boat from or carrying a burden over. Usually the
empty boat can be dropped down with a line, but
when the water is high boat as well as cargo must
be carried for 100 yards or more, and again, lower
down, for a less distance. The miners have put
down rollways along a roughly constructed road
here to make the portage of the boats easier, and
some windlasses for hauling the boats along the
water or out and into it. It would be possible to
build a good road or tramway along the east bank
of these rapids without great difficulty; and plans
are already formulated for a railway to be built
around the whole three miles of obstruction, in the
summer of 1898, to connect with the steamboats
above and below that will no doubt be running next
year.

The river below the rapids is fast (about four
miles an hour) for a few miles, and many gravel
banks appear. It gradually subsides, however, into
a quiet stream flowing northwest along the same
wide valley. No rock is seen here, the banks being
bluffs of white silt, which turns the clear blue of the
current above into a cloudy and opaque yellow.

Thirteen miles (measuring, as usual, along the river) brings the voyager to the mouth of the Tah-Keena, a turbid stream about 75 yards wide and 10 feet deep, which comes in from the west. Its sources are at the foot of the Chilkat Pass, where it flows out of West Kussoa lake (afterwards named Lake Arkell), and was formerly much employed by the Chilkat Indians as a means of reaching the interior, but was never in favor with the miners, and is now rarely followed by the Indians themselves, although its navigation from the lake down is reported to be easy.

Eleven and a half miles of quiet boating takes one to the head of Lake Labarge. This lake is 31 miles long, lies nearly north and south, and is irregularly elongated, reaching a width of six miles near the lower end. It is 2,100 feet above sea level and is bordered everywhere by mountains, those on the south having remarkably abrupt and castellated forms and carrying summits of white limestone. This lake is a very stormy one, and travelers often have to wait in camp for several days on its shores until calmer weather permits them to go on. This whole river valley is a great trough sucking inland the prevailing southerly summer winds, and navigation on all the lakes is likely to be rough for small boats.

The river below Lake Labarge is crooked, and at first rapid—six miles or more an hour, and interrupted by boulders; but it is believed that a stern wheel steamer of proper power could ascend at all times. The banks are earthen, but little worn, as floods do not seem to occur. Twenty-seven miles takes one to the mouth of a large tributary from the southeast,—the Teslintoo, which Schwatka called Newberry River, and which the miners mistakenly call Hotalinqu. It comes from the great Lake Teslin, which lies across the British Columbia boundary (Lat. 62 deg.), and is said to be 100 miles long; and it is further said that an Indian trail connects it with the head of canoe navigation on the Taku river, by only two long days of portaging. Some miners are said to have gone over it in 1876 or '77, Schwatka and Hayes came this way; and it may form one of the routes of the future,—perhaps even a railway route. This river flows through a wide and somewhat arid valley, and was roughly prospected about 1887 by men who reported finding fine gold all along its course, and also in tributaries of the lake. As the mountains about the head of the lake belong to the Cassiar range, upon whose southern slopes the Cassiar mines are situated, there is every reason to suppose that gold will ultimately be found there in paying quantities.

This part of the Lewes is called Thirty-mile River, under the impression that it is really a tributary of the Teslintoo, which is, in fact, wider than the Lewes at the junction (Teslintoo, width 575 feet; Lewes, 420 feet), but it carries far less water. From this confluence the course is north, in a deep, swift, somewhat turbid current, through the crooked defiles of the Seminow hills. Several auriferous bars have been worked here, and some shore-placers, including the rich Cassiar bar. Thirty-one miles below the Teslintoo the Big Salmon, or D'Abbadie River, enters from the southeast—an important river, 350 feet wide, having clear blue water flowing deep and quiet in a stream navigable by steamboats for many miles. Its head is about 150 miles away, not far from Teslin Lake, in some small lakes reached by the salmon, and surrounded by granite mountains. Prospectors have traced all its course and found fine gold in many places.

Thirty-four miles below the Big Salmon, west-north-west, along a comparatively straight course, carries the boatman to the Little Salmon, or Daly River, where the valley is so broad that no mountains are anywhere in sight, only lines of low hills at a distance from the banks. Five miles below this river the river makes an abrupt turn to the southwest around Eagle's Nest rock, and 18½ miles beyond

that reaches the Nordenskiold, a small, swift, clear-watered tributary from the southwest. The rocks of all this part of the river show thin seams of coal, and gold has been found on several bars. The current now flows nearly due north and a dozen miles below the Nordenskiold carries one to the second and last serious obstruction to navigation in the Rink rapids, as Schwatka called them, or Five-finger, as they are popularly known, referring to five large masses of rock that stand like towers in mid channel. These other islands back up the water and render its currents strong and turbulent, but will offer little opposition to a good steamboat. Boatmen descending the river are advised to hug the right bank, and a landing should be made twenty yards above the rapids in any eddy, where a heavily loaded boats should be lightened. The run should be made close along the shore, and all bad water ends when the Little Rink Rapids have been passed, six miles below. Just below the rapids the small Tatshun River comes in from the right. Then the valley broadens out, the current quiets down and a pleasing landscape greets the eye as bend after bend is turned. A long washed bank on the northeast side is called Hoo-che-koo Bluff, and soon after passing it one finds himself in the midst of the pretty Ingersoll archipelago, where the river widens out

and wanders among hundreds of islets. Fifty-five
miles by the river below Rink Rapids, the confluence
of the Lewes and Pelly is reached, and the first sign
of civilization in the ruins of old Fort Selkirk, with
such recent and probably temporary occupation as
circumstances may cause. Before long, undoubtedly, a flourishing permanent settlement will grow
up in this favorable situation.

The confluence here of the Lewes and Pelly rivers
forms the Yukon, which thenceforth pursues an un-
interrupted course of 1,650 miles to Behring Sea.
The country about the confluence is low, with extensive terrace flats running back to the bases of
rounded hills and ridges. The Yukon below the
junction averages about one-quarter of a mile in
width, and has an average depth of about 10 feet,
with a surface velocity of $4\frac{3}{4}$ miles an hour. A
good many gravel bars occur, but no shifting sand.
The general course nearly to White River, 96 miles,
is a little north of west, and many islands are seen;
then the river turns to a nearly due north course,
maintained at Fort Reliance. The White River is
a powerful stream, plunging down loaded with silt,
over ever shifting sand bars. Its upper source is
problematical, but is probably in the Alaskan
Mountains near the head of the Tenana and Forty-
mile Creek.

For the next ten miles the river spreads out to more than a mile wide and becomes a maze of islands and bars, the main channel being along the western shore, where there is plenty of water. This brings one to Stewart river, which is the most important right-hand tributary between the Pelly and the Porcupine. It enters from the east in the middle of a wide valley, and half a mile above its mouth is 200 yards in width; the current is slow and the water dark colored. It has been followed to its headquarters in the main range of the Rockies, and several large branches, on some of which there are remarkable falls, have been traced to their sources through the forested and snowy hills where they rise. These sources are perhaps 200 miles from the mouth, but as none of the wanderers were equipped with either geographical knowledge or instruments nothing definite is known. Reports of traces of precious metals have been brought back from many points in the Stewart valley, but this information is as vague as the other thus far. All reports agree that a light draught steamboat could go to the head of the Stewart and bar up its feeders. There is a trading post at its mouth.

The succeeding 125 miles holds what is at present the most interesting and populous part of the Yukon valley. The river varies from half to three-quarters of a mile wide and is full of islands. About 23 miles below Stewart River a large stream enters from the west called Sixty-mile Creek by the miners, who have had a small winter camp and trading store there for some years, and have explored its course for gold to its rise in the mountains west of the international boundary. Every little tributary has been named, among them (going up), Charley's Fork, Edwards Creek and Hawley Creek, in Canada, and then, on the American side of the line, Gold Creek, Miller Creek and Bed Rock Creek. The sand and gravel of all these have yielded fine gold and some of them, as Miller Creek, have become noted for their richness. Forty-four miles below Sixty-mile takes one to Dawson City, at the mouth of Klondike River, —the center of the highest productiveness and greatest excitement during 1897, when the gold fields of the interior of Alaska first attracted the attention of the world. Leaving to another special chapter an account of them, the itinerary may be completed by saying that $6\frac{1}{2}$ miles below the mouth of the Klondike is Fort Reliance, an old private trading post of no present importance. Twelve and a half miles farther the Chan-din-du River enters from the east, and $33\frac{1}{2}$ below that in the mouth of Forty-mile Creek, or Cone Hill River, which until the past year was the most important mining region of the inte-

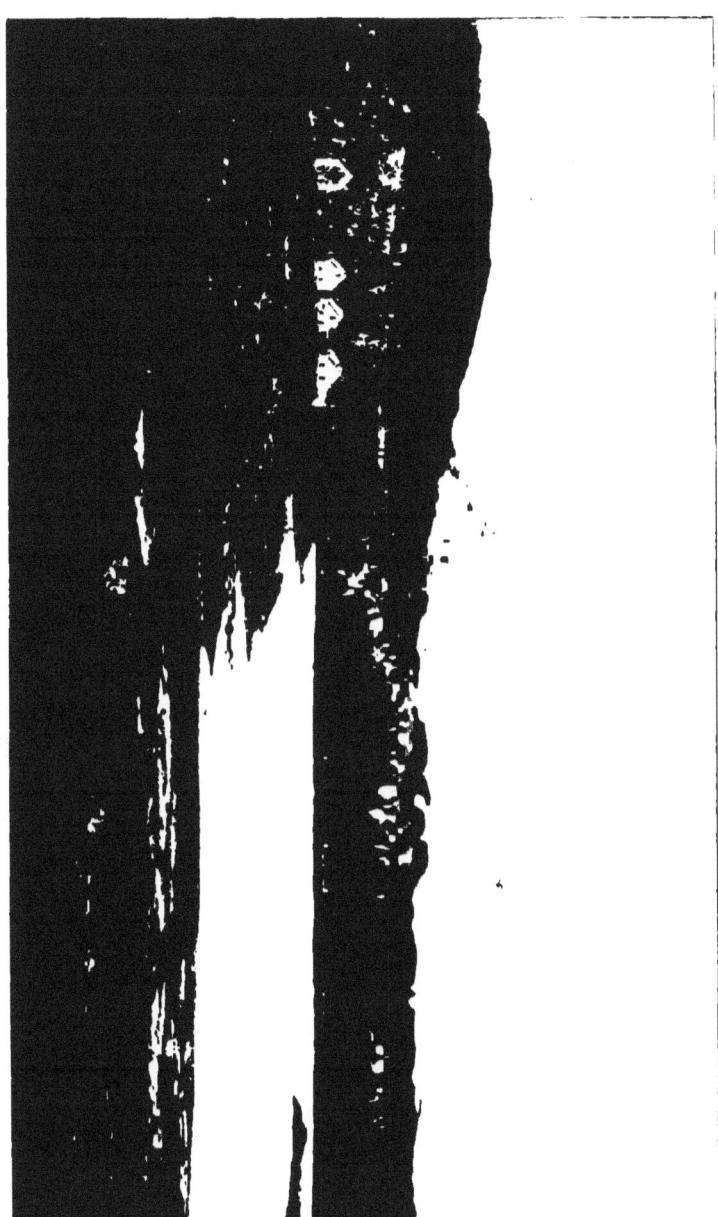

VIEW FROM INDIAN CHURCH, LOOKING NORTHEAST.

rior. It took its name from the supposition that it
was 40 miles from Fort Reliance, but the true distance is 46 miles. On the south side of the outlet of
this stream is the old trading post and modern town
of Forty-Mile, and on the north side the more recent settlement Cudahy. Both towns are, of course,
on the western bank of the Yukon, which is here
about half a mile wide. Five miles below Cudahy,
Coal Creek comes in from the east, and nearly
marks the Alaskan boundary, where a narrowed
part of the river admits one to United States territory. Prominent landmarks here are two great
rocks, named by old timers Old Man rock, on the
west bank, and Old Woman, on the east bank, in
reference to Indian legends attached to them. Some
twenty miles west of the boundary—the river now
having turned nearly due west in its general course
—Seventy-mile, or Klevande Creek, comes in from
the south, and somewhat below it the Tat-on-duc
from the north. It was ascended in 1887 by Mr.
Ogilvie, who describes its lower valley as broad and
well timbered, but its upper part flows through a
series of magnificent cañons, one of which half a
mile long, is not more than 50 feet wide with vertical
walls fully 700 feet in height. There are said to be
warm sulphur springs along its course, and the Indians regard it as one of the best hunting fields,

sheep being especially numerous on the mountains
in which it heads, close by the international boundary, where it is separated by only a narrow divide
from Ogilvie River, one of the head streams of the
Peel river, and also from the head of the Porcupine,
to which there is an Indian trail. Hence the miners
call this Sheep River. The rocks along this stream
are all sandstones, limestone and conglomerates,
with many thin calcite veins. Large and dense timber prevails, and game is abundant.

Below the mouth of the Tat-on-duc several small
streams enter, of which the Kandik on the north
and the Kolto or Charley's River—at the mouth of
which there used to be the home of an old Indian
notability named Charley—are most important.
About 160 miles from the boundary the Yukon flats
are reached, and the center of another important
mining district—that of Birch Creek and the Upper
Tenana—at Circle City, the usual terminus of the
trip up the Lower Yukon from St Michael.

HISTORY AND CHARACTERISTICS OF THE UPPER YUKON VALLEY.

The sources of the Yukon are just within the northern boundary of British Columbia (Lat. 62 deg.) among a mass of mountains forming a part of the great uplift of the Coast range, continuous with the Sierras of California and the Puget Sound coast. Here spring the sources of the Stikeen, flowing southwest to the Pacific, of the Fraser, flowing south through British Columbia, and of the Liard flowing northeasterly to the Mackenzie. Headwaters of the Stikeen and Liard interlock, indeed, along an extensive or sinuous watershed having an elevation of 3,000 feet or less and extending east and west. There are, however, many wide and comparatively level bottom lands scattered throughout this region and numerous lakes. The coast ranges here have an average width of about eighty miles and border the continent as far north as Lynn Canal, where they trend inland behind the St. Elias Alps. Many of their peaks exceed 8,000 feet in height, but few districts have been explored west. Eastward of this mountain axis, and separated from it by the valleys of the Fraser and Columbia in the south and the Yukon northward, is the Continental Divide, or Rocky Mountains proper, which is broken through (as noted above) by the Laird, but north of that cañon-bound river forms the watershed between the Liard and Yukon and between the Yukon and Mackenzie. These summits attain a height of 7,000 to 9,000 feet, and rise from a very complicated series of ranges extending northward to the Arctic Ocean, and very little explored. The valley of the Yukon, then, lies between the Rocky Mountains, separating its drainage basin from that of the Mackenzie, and the Coast range and St. Elias Alps separating it from the sea. Granite is the principal rock in both these great lines of watershed-uplift, and all the mountains show the effects of an extensive glaciation, and all the higher peaks still bear local remnants of the ancient ice-sheet.

The headwaters of the great river are gathered into three principal streams. First, the Lewes, easternmost, with its large tributaries, the Teslintoo and Big Salmon; second, the Pelly, with its great western tributary, the MacMillon.

The Lewes River has been described. It was known to the fur traders as early as 1840, and the Chilkat and Chilkoot passes were occasionally used by their Indian couriers from that time on. The gold fields in British Columbia from 1863 onwards

SCENE IN JUNEAU -- MOUNTAINS AND INDIAN HOUSES

stimulated prospecting in the northern and coastal parts of that province, and in 1872 prospectors reached the actual headwaters of the Lewes from the south, but were probably not aware of it; and that country was not scientifically examined until the reconnoisance of Dr. G. M. Dawson in 1887. In 1866 Ketchum and La Barge, of the Western Union Telegraph survey, ascended the Lewes as far as the lakes still called Ketchum and La Barge. In 1883 Lieut. Frederick Schwatka, U. S. A., and an assistant named Hayes, and several Indians, made their way across from Taka inlet to the head of Tahgish (a Tako) Lake, and descended the Lewes on a raft to Fort Selkirk, studying and naming the valley. From Fort Selkirk an entirely new route was followed toward the mountains forming the divide between the Yukon and the White and Copper rivers, which flow to the Gulf of Alaska, north of Mt. St. Elias. After discovering a pass little more than 5,000 feet high, they struck the Chityna River and followed that to the Copper River and thence to the coast. The Copper River Valley was thoroughly explored somewhat later by Lieuts. Abercrombie and Allen, U. S. A., who added greatly to knowledge of that large river, which, however, seems to have no good harbor at its mouth. The miners began to use the Chilkoot Pass and the Lewes River route to

the Yukon district in 1884. Some additions were
made to geography in this region by an exploring
expedition despatched to Alaska in 1890 by Frank
Leslie's Weekly, under Messrs. A. J. Wells, E. J.
Glave and A. B. Schanz. They entered by way of
Chilkat pass and came to a large lake at the head
of the Tah-keena tributary of the Lewes, which they
named Lake Arkell, though it was probably the
same earlier described by the Drs. Krause. Here
Mr. Glave left the party and striking across the coast
range southward discovered the headwaters of the
Alsekh and descended to Dry Bay. At Forty-mile
creek Mr. Wells and a party crossed over into the
basin of the Tanana and increased the knowledge of
that river. Mr. Schanz went down the Yukon
and explored the lower region. In 1892 Mr. Glave
again went to Alaska, demonstrated the possibility
of taking pack horses over the Chilkat trail, and
with an aid named Dalton made an extensive journey southward along the crest of the watershed between the Yukon valley and the coast.

Turning now to the Pelly, we find that this was
the earliest avenue of discovery. The Pelly rises in
lakes under the 62nd parallel, just over a divide from
the Finlayson and Frances Lake, the head of the
Frances River, the northern source of the Liard, and
this region was entered by the Hudson Bay Com-

pany as early as 1834, and gradually exploring the
Laird River and its tributaries, in 1840 Robert Cam-
bell crossed over the divide north of Lake Finlayson (at the head of the Frances), and discovered (at
a place called Pelly Banks) a large river flowing
northwest which he named Pelly. In 1843 he descended the river to its confluence with the Lewes
(which he then named), and in 1848 he built a post
for the H. B. Company at that point, calling it Fort
Selkirk. This done, in 1850, Campbell floated down
the river as far as the mouth of the Porcupine,
where three years previously (1847) Fort Yukon had
been established by Mr. Murray, who (founded by
James Bell in 1842) crossed over from the mouth
of the Mackenzie. The Yukon may thus be said
to have been "discovered" at several points inde-
pendently. The Russians, who knew it only at the
mouth, called it Kwikhpak, after an Eskimo name.
The English at Fort Yukon, learned that name from
the Indians there, and the upper river was the Pelly.
The English and Russian traders soon met, and
when Campbell came down in 1850 the identity of
the whole stream was established. The name Yukon gradually took the place of all others on English
maps and is now recognized for the whole stream
from the junction of the Lewes and Pelly to the
delta.

The Yukon basin, east of the Alaskan boundary, is known in Canada as the Yukon district, and contains about 150,000 square miles. This is nearly equal to the area of France, is greater than that of the United Kingdom of Great Britain and Ireland by 71,000 square miles, and nearly three times bigger than that of the New England states. To this must be added an area of about 180,000 square miles, west of the boundary, drained by the Yukon upon its way to the sea through Alaska. Nevertheless, Dr. G. M. Dawson and other students of the matter are of the opinion that the river does not discharge as much water as does the Mackenzie—nor could it be expected to do so, since the drainage area of the Mackenzie is more than double that of the Yukon, while the average annual precipitation of rain over the two areas seems to be substantially similar. Remembering these figures and that the basin of the Mississippi has no less than 1,225,000 square miles as compared with the 330,000 square miles of the Yukon basin, it is plain that the statement often heard that the Yukon is next to the Mississippi in size, is greatly exaggerated. In fact, its proportions, from all points of view, are exceeded by those of the Nile, Ganges, St. Lawrence and several other rivers of considerably less importance than the Mississippi.

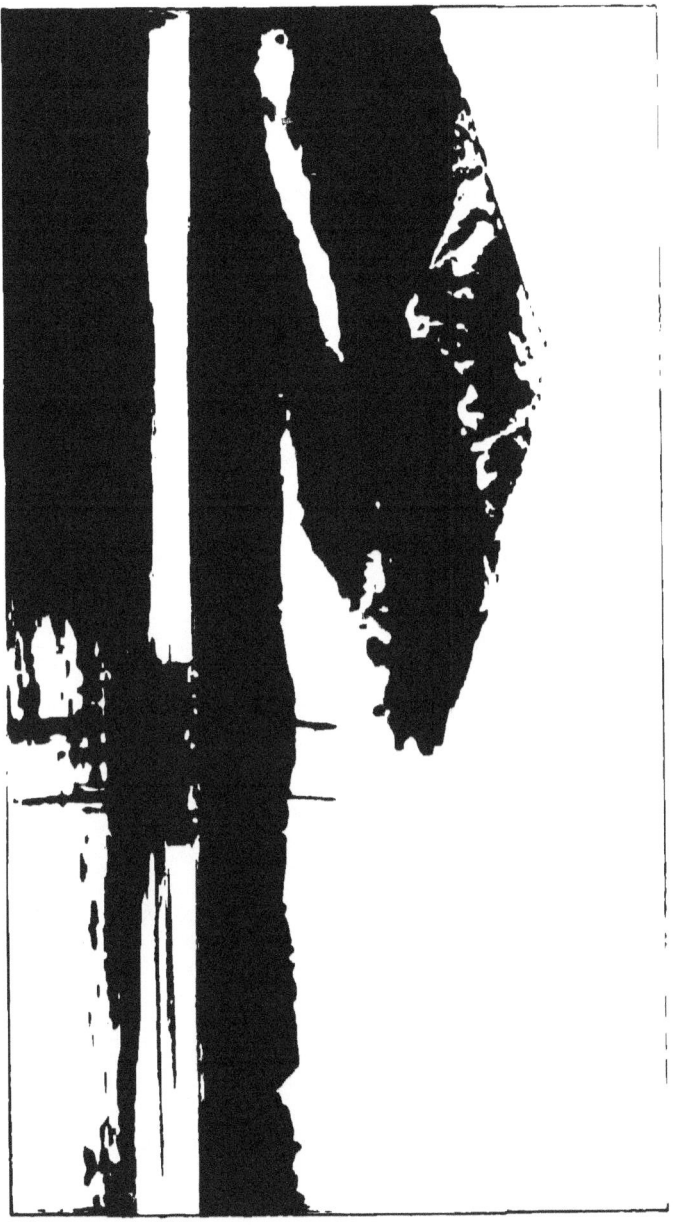

EARLY MORNING AT JUNEAU.

Resuming the historical outline, a short paragraph will suffice to complete the simple story down to the year 1896.

Robert Campbell had scarcely returned from his river voyage to his duties at Fort Selkirk, when he discovered that its location in the angle between the rivers was untenable, owing to ice-jams and floods. The station was therefore moved, in the season of 1852 across to the west bank of the Yukon, a short distance below the confluence, and new buildings were erected. These had scarcely been completely, when, on August 1st, a band of Chilkat Indians from the coast came down the river and early in the morning seized upon the post, surprising Mr. Campbell in bed, and ordered him to take his departure before night. They were not at all rough with him or his few men, but simply insisted that they depart, which they did, taking such personal luggage as they could put into a boat and starting down stream. The Indians then pillaged the place, and after feasting on all they could eat and appropriating what they could carry away, set fire to the remainder and burned the whole place to the ground. One chimney still stands to mark the spot, and others lie where they fell. This act was not dictated by wanton destructiveness on the part of the Chilkats—bad as they undoubtedly were and are; but was in pursuance of a theory. The establishment of the post there interfered with the monopoly of trade that they had enjoyed theretofore, with all the Indians of the interior, to whom they brought salable goods from the coast, taking in exchange furs, copper, etc., at an exhorbitant profit, which they enforced by their superior brutality. The Hudson Bay Company was robbing them of this, hence the demolition of the post, which was too remote to be profitably sustained against such opposition.

A little way down the river, Mr. Campbell met a fleet of boats bringing up his season's goods, and many friendly Indians. These were eager to pursue the robbers, but Campbell thought it best not to do so. He turned the supply-boats back to Fort Yukon and led his own men up the Pelly and over the pass to the Frances and so down the Liard to Fort Simpson, on the Mackenzie. Such is the story of the ruins of Fort Selkirk. Fort Yukon flourished as the only trading post until the purchase of Alaska by the United States, when Captain Raymond, an army officer, was sent to inform the factor there that his post was on United States territory, and require him to leave. He did so as soon as Rampart House could be built to take its place up the Porcupine. Old Fort Yukon then fell into ruins, and Rampart House itself was soon abandoned. In 1873 an opposition appeared in the independent trading house of Harper & McQuestion, men who had come into the country from the south; after long experience in the fur trade. They had posts at various points, occupied Fort Reliance for several years, and in 1886 established a post at the mouth of the Stewart River for the miners who had begun to gather there two years before. Many maps mark "Reed's House" as a point on the upper Stewart, but no such a trading-post ever existed there, although there was a fishing station and shelter-hut on one of its upper branches at an early day. This firm became the representatives of the Alaska Commercial Company (a San Francisco corporation) and opened a store in 1887 at Forty Mile, where they still do business.

Gold Discoveries.—The presence of fine float gold in river sands was early discovered by the Hudson Bay Company men, but in accordance with the former policy of that company, no mining was done and as little said about it as possible. The rich · ness of the Cassiar mines led to some prospecting northward as early as 1872, and by 1880 wandering gold hunters had penetrated to the Testintos, where for several years $8 to $10 a day of fine gold was sluiced out during the season by the small colony. In 1886 Cassiar Bar, on the Lewes, below there, was opened, and a party of four took out

$6,000 in 30 days, while other neighboring bars
yielded fair wages.   By that time Stewart River was
becoming attractive, and many miners worked placers there profitably in 1885, '86 and '87.   During
the fall of 1886 three or four men took the engines
out of the little steamboat "New Racket," which
was laid up for the winter there, and used them to
drive a set of pumps lifting water into sluice-boxes;
and with this crude machinery each man cleared
$1,000 in less than a month.   A judicious estimate
is, that the Stewart River placers yielded $100,000
in 1885 and '86.

Prospecting went on unremittingly, but nothing
else was found of promise until 1886, when coarse
gold was reported upon Forty Mile Creek, or the
Shitando River, as it was known to the Indians, and
a local rush took place to its cañons, the principal
attraction being Franklin Gulch, named after its dis-
coverer.   Three or four hundred men gathered there
by the season of 1887, and all did well.   This stream
is a "bed-rock" creek,—that is, one in the bed of
which there is very little drift; and in many places
the bed-rock was scraped with knives to get the little loose stuff out of crannies.   Some nuggets were
found.   At its mouth are extensive bars along the
Yukon, which carry gold throughout their depth.
During 1888 the season was very unfavorable and

HARBOR OF SITKA.

not much accomplished. Sixty Mile Creek was brought to notice, and Miller Gulch proved richer than usual. It is one of the headwaters of Sixty Mile, and some 70 miles from the mouth of the river where, in 1892, a trading store, saw-mill and little wintering-town was begun. Miller Creek is about 7 miles long, and its valley is filled with vast deposits of auriferous drift. In 1892 rich strikes were made and 125 miners gathered there, paying $10 a day for help, and many making fortunes. One clean-up of 1,100 ounces was reported. Glacier Creek, a neighboring stream, exhibited equal chances and drew many claimants, some of whom migrated thither in mid-winter, drawing their sleds through the woods and rocks with the murcury 30 degrees below zero. All of these gulches and other golden headwaters on both Forty Mile and Sixty Mile Creek, are west of the boundary in Alaska; but the mouths of the main streams and supply points are in Canadian territory. In all, the great obstacle is the difficulty of getting water up on the bars without expensive machinery; and the same is true of the rich gravel along the banks of the Yukon itself. Birch Creek was the next find of importance, and was promising enough to draw the larger part of the local population, which by this time had been considerably increased, for the news of the rich-

ness of the Forty Mile gulches had reached the outside world and attracted adventurous men and not
a few women from the coast not only, but from
British Columbia and the United States. A rival to
Harper & McQuestion, agents of the Alaska Commercial Company, appeared in the North American
Transportation and Trading Company, which increased the transportation service on the Yukon
River, by which most of the new arrivals entered,
and by establishing large competitive stores at Fort
Cudahy (Forty Mile) and elsewhere reduced the
price of food and other necessaries. About this
time, also, the Canadian government sent law officers and a detachment of mounted police, so that
the Yukon District began to take a recognized place
in the world.

Birch Creek is really a large river rising in the
Iauana Hills, just west of the boundary and flowing
northwest, parallel with the Yukon, to a debouch-
ment some 20 miles west of Fort Yukon. Between
the two rivers lie the "Yukon Flats," and at one
point they are separated by only six miles. Here,
at the Yukon end of the road arose Circle City, so-
called from its proximity to the Arctic Circle. This
is an orderly little town of regular streets, and has
a recorder of claims, a store, etc.

Birch Creek has been thoroughly explored, and

in 1894 yielded good results. The gold was in coarse flakes and nuggets, so that $40 a day was made by some men, while all did well. The drift is not as deep here as in most other streams, and water can be applied more easily and copiously,— a vast advantage. Molymute, Crooked, Independence, Mastadon and Preacher creeks are the most noteworthy tributaries of this rich field.

The Koyukuk River, which flows from the borders of the Arctic Ocean, gathering many mountain tributaries, to enter the Yukon at Nulato, was also prospected in 1892, '93 and '94, and indications of good placers have been discovered there, but the northerly, exposed and remote situation has caused them to receive little attention thus far.

THE KLONDIKE.

During the autumn of 1896 several men and wo·
men, none of whom were "old miners," discouraged
by poor results lower down the river resolved to
try prospecting in the Klondike gulch. They were
laughed at and argued with; were told that prospectors years ago had been all over that valley, and
found only the despised "flour gold," which was too
fine to pay for washing it out. Nevertheless they
persisted and went at work. Only a short time
elapsed, when, on one of the lower southside
branches of the stream they found pockets of flakes
and nuggets of gold far richer than anything Alaska
had ever shown before. They named the stream
Bonanza, and a small tributary El Dorado. Others
came and nearly everyone succeeded. Before spring
nearly a ton and a half of gold had been taken from
the frozen ground. Nuggets weighing a pound
(troy) were found. A thousand dollars a day was
sometimes saved despite the rudeness of the methods,
but these things happened where pockets were
struck. Probably the total clean-up from January
to June was not less than $1,500,000. The report

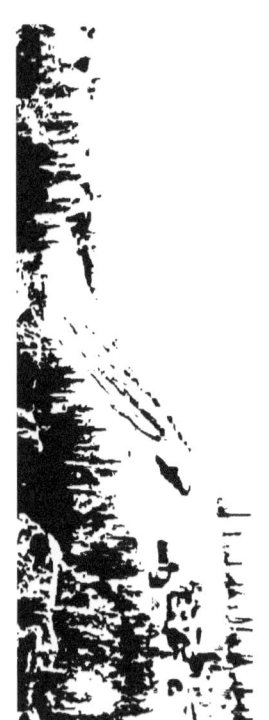

spread and all those in the interior of Alaska concentrated there, where a "camp" of tents and shanties soon sprang up at the mouth of the Klondike called Dawson City. A correspondent of the New York Sun describes it as beautifully situated, and a very quiet, orderly town, due to the strict supervision of the Canadian mounted police, who allowed no pistols to be carried, but a great place for gambling with high stakes. It bids fair to become the mining metropolis of the northwest, and had about 3,000 inhabitants before the advance-guard of the present "rush" reached there.

Hundreds of claims were staked out and worked in all the little gulches opening along Bonanza, Eldorado, Hunker, Bear and other tributaries of the Klondike, and of Indian River, a stream thirty miles south of it, and a greater number seem to be of equal richness with those first worked. All this is within a radius south and east of 20 miles from Dawson City, and most of it far nearer. The country is rough, wooded hills, and the same trouble as to water is met there as elsewhere, yet riches were obtained by many men in a few weeks without exhausting their claims.

So remote and shut in has this region been in the winter that no word of this leaked out until the river opened and a party of successful miners came down to the coast and took passage on the steamer Excelsior for San Francisco. They arrived on July 14, and no one suspected that there was anything extraordinary in the passenger list or cargo, until a procession of weather beaten men began a march to the Selby Smelting works, and there began to open sacks of dust and nuggets, until the heap made something not seen in San Francisco since the days of '49. The news flashed over the world, and aroused a fire of interest; and when three days later the Portland came into Seattle, bringing other miners and over $1,000,000 in gold, there was a rush to go north which bids fair to continue for months to come, for one of the articles of faith in the creed of the Yukon miner is that many other gulches will be found as rich as these. One elderly man, who went in late last fall and with partners took four claims on Eldorado Creek, told a reporter that his pickings had amounted to $112,000, and that he was confident that the ground left was worth $2,000,000 more. "I want to say," he exclaims, "that I believe there is gold in every creek in Alaska. Certain on the Klondike the claims are not spotted. One seems to be as good as another. It's gold, gold, gold, all over. It's yards wide and deep. All you have to do is to run a hole down."

One might go on quoting such rhapsodies, arising from success, to end of the book, but it is needless, for every newspaper has been full of them for a month.

One man and his wife got $135,000; another, formerly a steamboat deck-hand, $150,000; another, $115,000; a score or more over $50,000, and so on. These sums were savings after having the heavy expenses of the winter, and most of them had dug out only a small part of their ground.

It is curious in view of this success to read the only descriptive note the present writer can discover in early writings as to this gold river. It occurs in Ogilvie's report of his explorations of 1887, and is as follows: "Six and a half miles above Reliance the Tou-Dac River of the Indians (Deer River of Schwatka) enter from the east. It is a small river about 40 yards wide at the mouth and shallow; the water is clear and transparent and of a beautiful blue color. The Indians catch great numbers of salmon here. A miner had prospected up this river for an estimated distance of 40 miles in the season of 1887. I did not see him."

THE METHODS OF PLACER MINING

in the Klondike region and elsewhere along the Yukon are different from those pursued elsewhere, owing to the fact that from a point about three feet below the surface the ground is permanently frozen. The early men tried to strip off the gravel down to the gold lying in its lower levels or beneath it, upon the bed rock, and found it exceedingly slow and laborious work; moreover, it was only during the short summer that any work could be done. Now, by the aid of fires they sink shafts and then tunnel along the bed rock where the gold lies. A returned miner described the process as follows, pointing out the great advantage of being able to work under ground during the winter:

"The miners build fires over the area where they wish to work and keep these lighted over that territory for the space of twenty-four hours. Then the gravel will be melted and softened to a depth of perhaps six inches. This is then taken off and other fires are built until the gold bearing layer is reached. When the shaft is down that far other fires are built at the bottom, against the sides of the layer and tunnels made in the same manner. Blasting will do

PLACER MINE, CLAIM No. 3, ON MILLER CREEK.

no good, the charge not cracking off but blowing out of the hole. The matter taken out, and containing the gold is piled up until spring, when the torrents come down, and is panned and cradled by these. It is certainly very hard labor."

Another quotation may be given as a practical example of this process:

"The gold so far as has been taken from Bonanza and Eldorado, both well named, for the richness of the placers are truly marvelous. Eldorado, thirty miles long, is staked the whole length and as far as worked has paid.

"One of our passengers, who is taking home $100,000 with him, has worked one hundred feet of his ground and refused $200,000 for the remainder, and confidently expects to clean up $400,000 and more. He has in a bottle $212 from one pan of dirt. His pay dirt while being washed averaged $250 an hour to each man shoveling in. Two others of our miners who worked their own claim cleaned up $6,000 from one day's washing.

"There is about fifteen feet of dirt above bed rock, the pay streak averaging from four to six feet, which is tunnelled out while the ground is frozen. Of course, the ground taken out is thawed by building fires, and when the thaw comes and water rushes in they set their sluices and wash the dirt. Two of our fellows thought a small bird in the hand worth a large one in the bush, and sold their claims for $45,000, getting $4,500 down, and the remainder to be paid in monthly installments of $10,000 each. The purchasers had no more than $5,000 paid. They were twenty days thawing and getting out dirt. Then there was no water to sluice with, but one fellow made a rocker, and in ten days took out the $10,000 for the first installment. So, tunnelling and rocking, they took out $40,000 before there was water to sluice with."

LEGAL ASPECT OF ALASKA.

Commissioner Hermann, of the General Land Office, has announced that the following laws of the United States extend over Alaska, where the general land laws do not apply:

First—The mineral land laws of the United States.

Second—Town-site laws, which provide for the incorporation of town-sites and acquirement of title thereto from the United States Government by the town-site trustees.

Third—The laws providing for trade and manu-factures, giving each qualified person 160 acres of land in a square and compact form.

The coal land regulations are distinct from the mineral regulations or laws, and as in the case of the general land laws Alaska is expressly exempt from this jurisdiction.

On the part of Canada, however, the provisions of the Real Property act of the Northwest Territories will be extended to the Yukon country by an order in council, a register will be appointed, and a land title office will be established.

The act approved May 17, 1884, providing a civil government for Alaska, has this language as to mines and mining privileges:

"The laws of the United States relating to min-ing claims and rights incidental thereto shall, on and after the passage of this act, be in full force and effect in said district of Alaska, subject to such regulations as may be made by the Secretary of the Interior and approved by the President," and "parties who have located mines or mining priv-ileges therein, under the United States laws ap-plicable to the public domain, or have occupied or improved or exercised acts of ownership over such claims, shall not be disturbed therein, but shall be allowed to perfect title by payment so provided for."

There is still more general authority. Without
the special authority, the act of July 4. 1866, says:
"All valuable mineral deposits in lands belonging to
the United States, both surveyed and unsurveyed,
are hereby declared to be free and open to exploration and purchase, and lands in which these are
found to occupation and purchase by citizens of the
United States and by those who have declared an
intention to become such, under the rules prescribed
by law and according to local customs or rules of
miners in the several mining districts, so far as the
same are applicable and not inconsistent with the
laws of the United States."

The patenting of mineral lands in Alaska is not
a new thing, for that work has been going on, as
the cases have come in from time to time, since
1884.

One of the difficulties that local capitalists find in
their negotiations for purchase of mining properties
on the Yukon is the lack of authenticated records
of owners of claims. Different practices prevail on
the two sides of the line and cause more or less con-
fusion. The practice has been at most of the new
camps to call a miners' meeting at which one of the
parties was elected recorder, and he proceeded to
enter the bearings of stakes and natural marks to
define claims. Sometimes the recorder would give

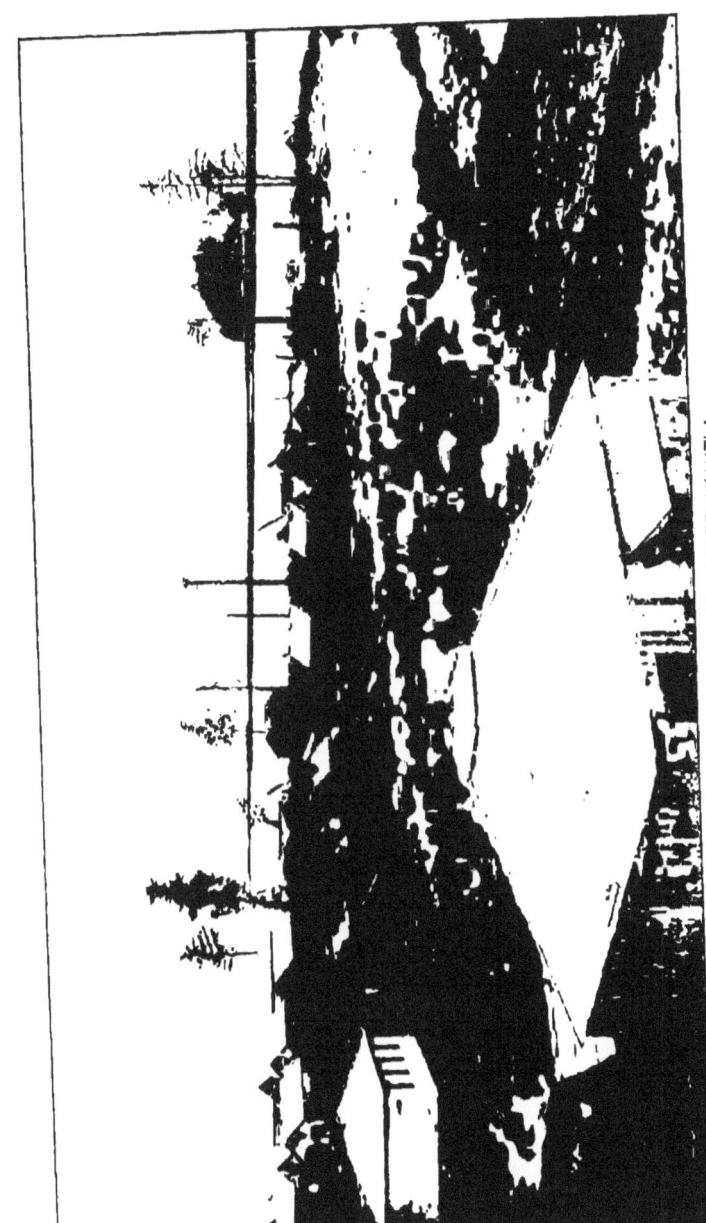

THE POINT AND BEACH AT METLAKAHTLA.

a receipt for a fee allowed by common consent for
recording, and also keep a copy for future reference,
but in a majority of cases even this formality was
dispensed with, and the only record kept was the
rough minutes made at the time.

On the Canadian side a different state of affairs
exists. The Dominion Government has sent a commissioner who is empowered to report officially all
claims, and while no certificate is issued to the owners thereof, properties are thoroughly defined and
their metes and bounds established. The commissioner in the Klondike district, whose name is Constantine, also exercises semi-judicial functions, and
settles disputes to the best of his ability, appeal lying
to the Ottawa Government.

As to courts and the execution of civil and criminal
law generally, none were existent in the upper Yukon Valley on the American side of the line during
1897. The nearest United States judge was at Sitka. At Circle City and other centers of population
the people had organized into a sort of town-meet-
ing for the few public matters required; and a sort
of vigilance committee took the place of constituted
authority and police. As a matter of fact, however,
the people were quiet and law-abiding and little need
for the machinery of law is likely to arise before
courts, etc., are set up. A movement toward send-

ing a garrison of United States troops thither was
vetoed by the War Department.

Canada, however, awoke to the realization that
her interests were in jeopardy, and took early steps
to profit by the wealth which had been discovered
within her borders and the international business
that resulted.   The natural feeling among the Canadians was, and is, that the property belongs to
the Canadian public, and that no good reason exists
why the mineral and other wealth should be ex·
hausted at once, mainly by outsiders, as has largely
happened in the case of Canada's forests. A pro-
hibitory policy was urged by some, but this seemed
neither wise nor practicable; and the Dominion
Government set at work to save as large a share as
it could.   As there are gold fields on the Alaska
side of the line, and the approaches lie through
United States territory, a spirit of reciprocal accommodation was necessary.   One difficulty has been
averted last spring by President Cleveland's veto
of the Immigration bill, one provision of which
would have prevented Canadian laborers drawing
wages in this country, and probably would have pro-
voked a retaliatory act.

Canada has already placed customs officers on the
passes and at the Yukon crossing of the boundary
to collect customs duties not only on merchandise

but on miner's personal outfits. There is practically
no exception, and the duty comes below 20 per cent.
on but few articles. On most of the goods the duty
is from 30 to 35 per cent., and in several instances
higher, but the matter may be very simply ad-
justed by purchasing tools and outfits in Victoria
or Vancouver, for thus far the United States has
placed no corresponding obstruction in the way of
Canadian travellers to the gold-fields, but, on the
contrary, has made Dyea a sub-port of entry, largely
to accommodate British transportation lines. The
Canadian Government is represented in that region
now only by customs officers and 20 mounted police,
but it is taking steps to garrison the whole upper
Yukon Valley with its mounted police,—a body of
officers, whose functions are half military, half civil,
and which, it may as well be conceded once for all,
cannot be trifled with. There is no question but
that they will do their level best to enforce the laws
to the utmost. The commander of each detachment
will be constituted a magistrate of limited powers,
so that civil examinations and trials may be speedily
conducted.

The plan is to erect a strong post a short distance
north of the sixtieth degree of latitude, just above
the northern boundary of British Columbia, and beyond the head of the Lynn Canal, where the Chil-

koot Pass and the White Pass converge. This post
will command the southern entrance to the whole
of that territory. Further on small police posts will
be established, about fifty miles apart, down to Fort
Selkirk, while another general post will patrol the
river near the international boundary, with headquarters, probably, in the Klondike valley.

The mining regulations of Canada, applying to
the Yukon placer claims, are as follows:

"Bar diggings" shall mean any part of a river
over which water extends when the water is in its
flooded state and which is not covered at low water.
"Mines on benches" shall be known as bench diggings, and shall for the purpose of defining the size
of such claims be excepted from dry diggings. "Dry
diggings" shall mean any mine over which a river
never extends. "Miner" shall mean a male or fe-
male over the age of eighteen, but not under that
age. "Claims" shall mean the personal right of
property in a placer mine or diggings during the
time for which the grant of such mine or diggings
is made. "Legal post" shall mean a stake standing
not less than four feet above the ground and squared
on four sides for at least one foot from the top.
"Close season" shall mean the period of the year
during which placer mining is generally suspended.
The period to be fixed by the gold commissioner

FORT WRANGELL.

in whose district the claim is situated. "Locality" shall mean the territory along a river (tributary of the Yukon) and its affluents. "Mineral" shall include all minerals whatsoever other than coal.

1. Bar diggings. A strip of land 100 feet wide at highwater mark and thence extending along the river to its lowest water level.

2. The sides of a claim for bar diggings shall be two parallel lines run as nearly as possible at right angles to the stream, and shall be marked by four legal posts, one at each end of the claim at or about high water mark; also one at each end of the claim at or about the edge of the water. One of the posts shall be legibly marked with the name of the miner and the date upon which the claim is staked.

3. Dry diggings shall be 100 feet square and shall have placed at each of its four corners a legal post, upon one of which shall be legibly marked the name of the miner and the date upon the claim was staked.

4. Creek and river claims shall be 500 feet long, measured in the direction of the mineral course of the stream, and shall extend in width from base to base of the hill or bench on each side, but when the hills or benches are less than 100 feet apart the claim may be 100 feet in depth. The sides of a claim shall be two parallel lines run as nearly as

possible at right angles to the stream. The sides shall be marked with legal posts at or about the edge of the water and at the rear boundary of the claim. One of the legal posts at the stream shall be legibly marked with the name of the miner and the date upon which the claim was staked.

5. Bench claims shall be 100 feet square.

6. In defining the size of claims they shall be measured horizontally, irrespective of inequalities on the surface of the ground.

7. If any person or persons shall discover a new mine and such discovery shall be established to the satisfaction of the gold commissioner, a claim for the bar diggings 750 feet in length may be granted. A new stratum of auriferous earth or gravel situated in a locality where the claims are abandoned shall for this purpose be deemed a new mine, although the same locality shall have previously been worked at a different level.

8. The forms of application for a grant for placer mining and the grant of the same shall be according to those made, provided or supplied by the gold commissioner.

9. A claim shall be recorded with the gold commissioner in whose district it is situated within three days after the location thereof if it is located within ten miles of the commissioner's office. One day

extra shall be allowed for making such record for every additional ten miles and fraction thereof

10. In the event of the absence of the gold commissioner from his office for entry a claim may be granted by any person whom he may appoint to perform his duties in his absence.

11. Entry shall not be granted for a claim which has not been staked by the applicant in person in the manner specified in these resolutions. An affidavit that the claim was staked out by the applicant shall be embodied in the application.

12. An entry fee of $15 shall be charged the first year and an annual fee of $100 for each of the following years:

13. After recording a claim the removal of any post by the holder thereof or any person acting in his behalf for the purpose of changing the boundaries of his claim shall act as a forfeiture of the claim.

14. The entry of every holder for a grant for placer mining must be renewed and his receipt relinquished and replaced every year, the entry fee being paid each year.

15. No miner shall receive a grant for more than one mining claim in the same locality; but the same miner may hold any number of claims by purchase, and any number of miners may unite to work their

claims in common upon such terms as they may arrange, provided such agreement be registered with the Gold Commissioner and a fee of $15 for each registration.

16. And miner may sell, mortgage, or dispose of his claims, provided such disposal be registered with and a fee of $5 paid to the Gold Commissioner.

17. Every miner shall, during the continuance of his grant, have the exclusive right of entry upon his own claim for the miner-like working thereof, and the construction of a residence thereon, and shall be entitled exclusively to all the proceeds realized therefrom; but he shall have no surface rights therein, and the Gold Commissioner may grant to the holders of adjacent claims such rights of entry thereon as may be absolutely necessary for the working of their claims, upon such terms as may to him seem reasonable. He may also grant permits to miners to cut timber thereon for their own use, upon payment of the dues prescribed by the regulation in that behalf.

18. Every miner shall be entitled to the use of so much of the water naturally flowing through or past his claim, and not already lawfully appropriated as shall in the opinion of the Gold Commissioner be necessary for the due working thereof, and shall be entitled to drain his own claim free of charge.

CHILKOOT PASS.

19. A claim shall be deemed to be abandoned and open to occupation and entry by any person when the same shall have remained unworked on working days by the guarantee thereof or by some person in his behalf for the space of seventy-two hours unless sickness or some other reasonable cause may be shown to the satisfaction of the Gold Commissioner, or unless the guarantee is absent on leave given by the commissioner, and the Gold Commissioner, upon obtaining satisfactory evidence that this provision is not being complied with, may cancel the entry given in the claim.

20. If the land upon which a claim has been located is not the property of the Crown it will be necessary for the person who applies for entry to furnish proof that he has acquired from the owner of the land the surface right before entry can be granted.

21. If the occupier of the lands has not received a patent thereof the purchase money of the surface rights must be paid to the Crown and a patent of the surface rights will issue to the party who acquired the mining rights. The money so collected will either be refunded to the occupier of the land when he is entitled to a patent there or will be credited to him on account of payment of land.

22. When the party obtaining the mining rights

cannot make an arrangement with the owner thereof for the acquisition of the surface rights it shall be
lawful for him to give notice to the owner or his
agents or the occupier to appoint an arbitrator to
act with another arbitrator named by him in order
to award the amount of compensation to which the
owner or occupier shall be entitled.

The royalty and reserve additions to this, made
since the recent discoveries and on account of them,
are as follows:

1. A royalty of 10 per cent will be collected for
the government on all amounts taken out of any one
claim up to $500 a week, and after that 20 per cent.
This royalty will be collected on gold taken from
streams already being worked, but in regard to all
future discoveries the government proposes

2. That upon every river and creek where mining
locations shall be staked out every alternate claim
shall be the property of the government.

These regulations, say the Canadians, are made
with the purpose of developing a country, which,
as elsewhere shown in this pamphlet, is capable of
supporting a large permanent population and varied
industries. Whether they can be enforced remains
to be seen, and difficulties will certainly attend the
collection of a royalty on gold-dust. The effect of
these regulations, it is believed by the authors, will

be to encourage permanent settlement and the treat-
ment of mining as a regular industry and not simply
as an adventurous speculation. Another effect, undoubtedly, will be to cause immigrants, including
Canadians themselves, to prospect and mine on the
United States side of the line, whenever they have
an equal opportunity for success.

The boundary dispute does not as yet seriously
affect the question or rights and privileges in the
new gold regions, as the disputed part of the line,
southeast of Alaska, runs through a region not yet
occupied, and practically the whole of Lynn Canal
is administered by the United States, and the Canadians act as though it were decided that their
boundary was farther inland than some of them
pretend. From Mt. St. Elias north, the 141st me-
ridian is the undisputed boundary, and this has been
fixed by an international commission, crossing the
Yukon at a marked point near the mouth of Forty
Mile Creek. Nearly or quite all of the diggings
upon which are written Alaskan territory, as also
are the valuable placers on Birch and Miller creeks.
It will be a matter of extreme difficulty along this
part of the boundary to prevent smuggling, to discover and collect Canadian royalties, and to capture
criminals except by international coöperation.

CLIMATE, AGRICULTURE AND HEALTH.

The Weather Bureau has made public a statement in regard to the climate of Alaska, which says:
"The climates of the coast and the interior of Alaska
are unlike in many respects, and the differences are
intensified in this as perhaps in few other countries
by exceptional physical conditions. The fringe of
islands that separates the mainland from the Pacific
Ocean from Dixon Sound north, and also a strip
of the mainland for possibly twenty miles back from
the sea, following the sweep of the coast as it curves
to the northwestward to the western extremity of
Alaska form a distinct climatic division which may
be termed temperate Alaska. The temperature rarely falls to zero; winter does not set in until Dec. 1,
and by the last of May the snow has disappeared except on the mountains.

"The mean winter temperature of Sitka is 32.5,
but little less than that of Washington, D. C. The
rainfall of temperate Alaska is notorious the world
over, not only as regards the quantity, but also as
to the manner of its falling, viz.: in long and incessant rains and drizzles. Cloud and fog naturally
abound, there being on an average but sixty-six
clear days in the year.

GENERAL VIEW OF SILVER BOW BASIN, NEAR JUNEAU.

"North of the Aleutian Islands the coast climate becomes more rigorous in winter, but in summer the difference is much less marked.

"The climate of the interior is one of extreme rigor in winter, with a brief but relatively hot summer, especially when the sky is free from cloud.

"In the Klondike region in midwinter the sun rises from 9:30 to 10 a. m., and sets from 2 to 3 p. m., the total length of daylight being about four hours. Remembering that the sun rises but a few degrees above the horizon and that it is wholly obscured on a great many days, the character of the winter months may easily be imagined.

"We are indebted to the United States coast and geodetic survey for a series of six months' observations on the Yukon, not far from the site of the present gold discoveries. The observations were made with standard instruments, and are wholly reliable. The mean temperatures of the months October, 1889, to April, 1890, both inclusive, are as follows: October, 33 degrees; November, 8 degrees; December, 11 degrees, below zero; January, 17 below zero; February, 15 below zero; March, 6 above; April 20 above. The daily mean temperature fell and remained below the freezing point (32 degrees) from Nov. 4, 1889, to April 21, 1890, thus giving 168 days as the length of the closed season of 1889-

'90, assuming that outdoor operations are controlled by temperature only. The lowest temperatures registered during the winter were: Thirty-two degrees below zero in November, 47 below in December, 59 below in January, 55 below in February, 45 below in March, and 26 below in April.

"The greatest continuous cold occurred in February, 1890, when the daily mean for five consecutive days was 47 degrees below zero.

"Greater cold than that here noted has been experienced in the United States for a very short time, but never has it continued so very cold for so long a time as in the interior of Alaska. The winter sets in as early as September, when snow-storms may be expected in the mountains and passes. Headway during one of those storms is impossible, and the traveler who is overtaken by one of them is indeed fortunate if he escapes with his life. Snow-storms of great severity may occur in any month from September to May, inclusive.

"The changes of temperature from winter to summer are rapid, owing to the great increase in the length of the day. In May the sun rises at about 3 a. m. and sets about 9 p. m. In June it rises about half past 1 in the morning, and sets at about half past 10, giving about twenty hours of daylight and diffuse twilight the remainder of the time.

"The mean summer temperature in the interior doubtless ranges between 60 and 70 degrees, according to elevation, being highest in the middle and lower Yukon valleys."

Accurate data of the temperature in the Klondike district were kept at Fort Constantine last year. The temperature first touched zero Nov. 10, and the zero weather recorded in the spring was on April 29.

Between Dec. 19 and Feb. 6 it never rose above zero. The lowest actual point, 65 below, accurred on Jan. 27, and on twenty-four days during the winter the temperature was below 50.

On March 12 it first rose above the freezing point, but no continuous mild weather occurred until May 4, after which date the temperature during the balance of the month frequently rose above 60 degrees.

The Yukon River froze up on Oct. 28 and broke up on May 17.

The long and severe winter and the frozen moss-covered ground are serious obstacles to agriculture and stock raising. The former can change but little with coming seasons, but the latter, by gradually burning off areas, can be overcome to some extent. On such burned tracts hardy vegetables have been and may be raised, and the area open to such use

is considerable. Potatoes do well and barley will
mature a fair crop.

Live stock may be kept by providing an abundance of shelter and feed and housing them during
the winter. In summer an abundance of the finest
grass pasture can be had, and great quantities of natural hay can be cut in various places.

Diseases: In spite of all that is heard in the
newspapers regarding the healthfulness of the climate of Alaska and the upper Yukon, the Census
Report of Alaska offers its incontestable statistics to
the effect that the country is not more salubrious,
nor its people more healthy than could be expected
in a region of violent climate, where the most ordinary laws of health remain almost totally ignored.
From the Government Report we quote the following:

"Those diseases which are most fatal to life in one
section of Alaska seem to be applicable to all others.
In the first place, the native children receive little
or no care, and for the first few years of their lives
are more often naked than clothed, at all seasons of
the year. Consumption is the simple and compre-
hensive title for the disease which destroys the
greater number of the people of Alaska. Aluet, Indian and Eskimo suffer from it alike; and all alike
exhibit the same stolid indifference to its slow and

MUIR GLACIER (MIDDLE PORTION).

fatal progress, make no attempt to ward it off, take no special precautions even when the disease reaches its climax.

Next to consumption, the scrofulous diseases, in the forms of ulcers, eat into the vitals and destroy them until the natives have the appearance of lepers to unaccustomed eyes. As a consequence of their neglect and the exigencies of the native life, forty or fifty years is counted among them as comparatively great age, and none are without the ophthalmic diseases necessarily attendant on existence in smoky barabaras. Against snow-blindness the Eskimo people use peculiar goggles, but by far the greater evil, the smoke poisoning of the ophmalmic nerve is neither overcome nor prevented by any of them. All traders carry medicine chests and do what they can to relieve suffering, but it requires a great deal of medicine to make an impression on the native constitution, doses being about four times what would suffice an Englishman or American.

OUTFITS, SUPPLIES, ETC.

Houses.—Almost every item has been taken into consideration by the prospectors starting out to face an Alaskan winter except the item of shelter when they shall have put their boats in winter dock. The result will be that many hundreds will find themselves in the bleak region with plenty of money and victuals, but insufficient protection from the cold weather. From accounts that have come from Alaska and British Columbia, there are more men there skilled in digging and bookkeeping than in carpentry, and more picks and shovels than axes and planes. With the arrival of parties that have lately gone to the headwaters of the Yukon, there will necessarily be an immense demand for houses, for without them the miners will freeze. This matter is beginning to receive attention in San Francisco and Seattle, and preparations are now under way to provide gold seekers with houses.

Within a week negotiations have been conducted between parties in San Francisco and this city for the shipment of entire houses to the gold regions. The houses will be constructed in sections, so that they may be carried easily in boats up the Yukon or packed on sleds and carried through the rough country in baggage trains. A New York

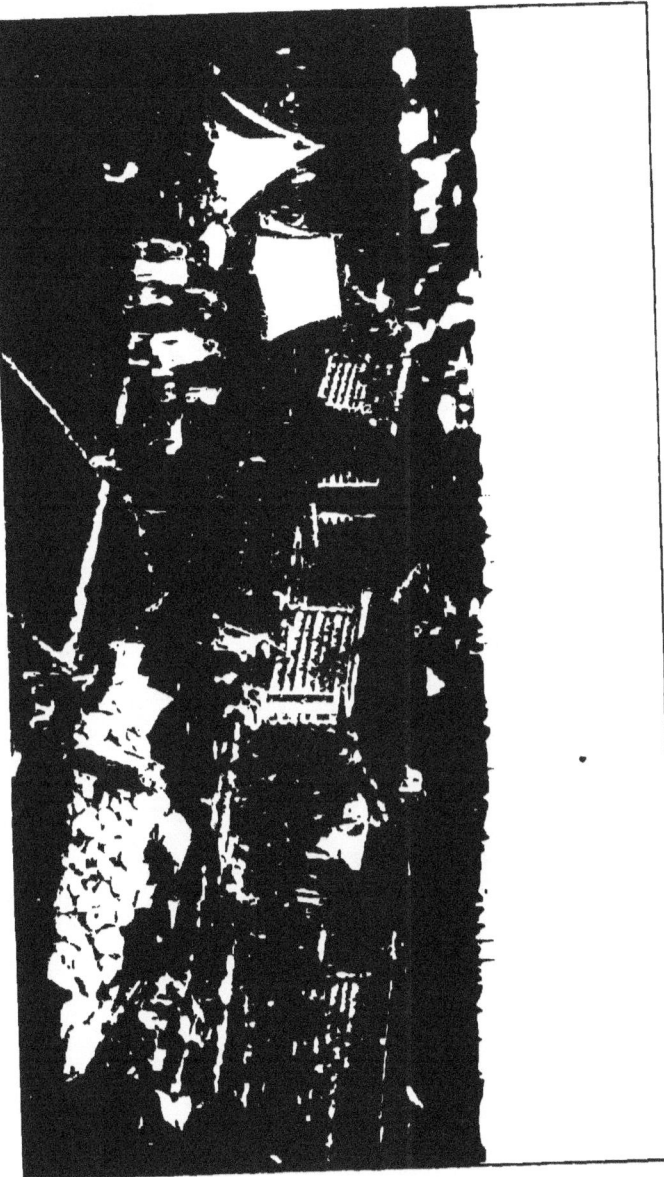

SUPPLY STATION FOR CIRCLE CITY.

firm which makes a specialty of such houses has received orders for as many as can be sent there.

No tents are used in winter, as they become coated with ice from the breath of the sleepers and are also apt to take fire.

Clothing for Men.—A year's supply of winter clothing ought be taken, especial pains being taken to supply plenty of warm, durable underwear. Old-timers in the country wear in winter a coat or blouse of dressed deer skin, with the hair on, coming down to the knees and held by a belt round the waist. It has a hood which may be thrown back on the shoulders when not needed. This shirt is trimmed with white deerskin or wolfskin, while those worn in extreme weather are often lined with fur. Next in importance to them are the torbassâ or Eskimo boots. These are of reindeer skin, taken from the legs, where the hair is short, smooth and stiff. These are sewed together to make the tops of the boots which come up nearly to the knee, where they are tied. The sole is of sealskin, turned over at heel and toe and gathered up so as to protect those parts and then brought up on each side. They are made much larger than the foot and are worn with a pad of dry grass which, folded to fit the sole, thickens the boot and forms an additional protection to the foot. A pair of strings tied about the ankle from either side complete a covering admirably adapted to the necessities of winter travel. If the newcomer can get such garments as these he will be well provided against winter rigors.

Women going to the mines are advised to take two pairs of extra heavy all-wool blankets, one small pillow, one fur robe, one warm shawl, one fur coat, easy fitting; three warm woollen dresses, with comfortable bodices and shirts knee length, flannel-lined preferable; three pairs of knickers or bloomers to match the dresses, three suits of heavy all-wool underwear, three warm flannel night dresses, four pairs of knitted woollen stockings, one pair of rubber boots, three gingham aprons that reach from neck to knees, small roll of flannel for insoles, wrapping the feet and bandages; a sewing kit, such toilet articles as are absolutely necessary, including some skin unguent to protect the face from the icy cold, two light blouses or shirt waists for summer wear, one oilskin blanket to wrap her effects in, to be secured at Juneau or St. Michael; one fur cape, two pairs of fur gloves, two pairs of surseal moccasins, two pairs of muclucs—wet weather moccasins.

She wears what she pleases en route to Juneau or St. Michael, and when she makes her start for the diggings she lays aside every civilized traveling garb, including shoes and stays, until she comes out.

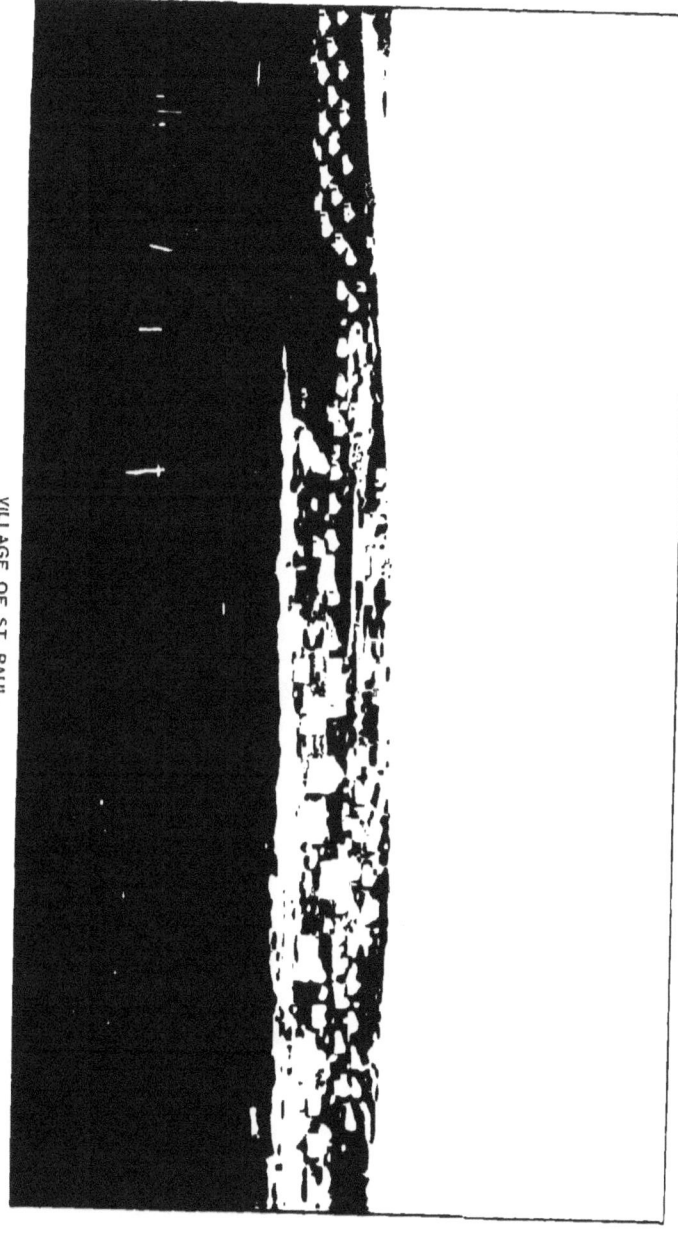

VILLAGE OF ST PAUL.

Instead of carrying the fur robe, fur coat and rubber boots along, she can get them on entering Alaska, but the experienced ones say, take them along. Leggings and shoes are not so safe nor desirable as the moccasins. A trunk is not the thing to transport baggage in. It is much better in a pack, with the oilskin cover well tied on. The things to add that are useful, but not absolutely necessary, are chocolate, coffee and the smaller light luxuries.

Beds are made on a platform raised a few feet from the floor, and about seven feet wide. Often consists of a raindeer skin with the hair on and one end sewn up so as to make a sort of bag to put the feet in. A pillow of wild goose feathers, and a pair of blankets. Sheets, which have been unknown heretofore, may become essential, but such a conventionality as a counterpane would better be left behind.

Provisions.—There was a report that Canadian mounted police would guard the passes during the latter part of the summer of 1897 and refuse admission to anyone who did not bring a year's provisions with him. This has been estimated as weighing 1,800 pounds. Whether this is true or not, it is certain that no one should go into the Yukon country without taking a large supply of food, and taking it from his starting-point. Whatever is the most

condensed and nutritious is the cheapest, and this
should be collected with great care. There is well-
grounded fear that famine may overtake all the
camps there before the opening of navigation in the
spring. Newspapers on August 2nd reported agents
of the Alaska Commercial Company as saying:

"We shall refuse to take passengers at all in our
next steamer. We could sell every berth at the
price we have been asking—$250, as against $120
last spring—but we shall not sell one. We shall
fill up with provisions, and I have no doubt the
Pacific Coast Company will do the same. We are
afraid. Those who are mad to get to the diggings
will probably be able to get transportation by char-
tering tramp steamers, and there is a serious risk
that there will not be food enough for them at Juneau or on the Yukon. After the season closes it
will be next to impossible to get supplies into the
Yukon country, and a large proportion of the gold
seekers may starve to death. That would be an
ominous beginning for the new camp. Alaska is
not like California or Australia or South Africa. It
produces nothing. When the supplies from outside are exhausted, famine must follow—to what degree no one can tell."

It was further understood at this date that there
are 2,000 tons of food at St. Michael, and the Alaska

PANORAMIC VIEW OF JUNEAU.

Company has three large and three small steamers to carry it up river. It is hard to ascertain how much there is at Juneau; it is vaguely stated that there are 5,000 tons. At a pinch steamers might work their way for several months to come through the ice to that port from Seattle, which is only three days distant. But it may be nip and tuck if there is any rush of gold seekers from the East.

Alaskan Mails.—Between Seattle and Sitka the mail steamers ply regularly. On the City of To-peka there has been established a regular sea post-office service. W. R. Curtis is the clerk in charge. Between Sitka and Juneau there is a closed pouch steamboat service. Seattle makes up closed pouches for Douglas, Fort Wrangel, Juneau, Killisnoo, Ket-chikan, Mary Island, Sitka, and Metlakatlah. Connecting at Sitka is other sea service between that point and Unalaska, 1,400 miles to the west. This service consists of one trip a month between Sitka and Unalaska from April to October and leaves Sit-ka immediately upon arrival of the mails from Seat-tle. Captain J. E. Hanson is acting clerk. From Unalaska the mails are dispatched to St. Michael and thence to points on the Yukon.

The Postoffice department has perfected not only a summer but a winter star route service between Juneau and Circle City. The route is overland and

by boats and rafts over the lakes and down the Yukon, and is 900 miles long.   A Chicago man named
Beddoe carries the summer mail, making five trips
between June and November, and is paid $500 a
trip.   Two Juneau men, Frank Corwin and Albert
Hayes, operate the winter service and draw for each
round trip $1,700 in gold.   About 1,200 letters are
carried on each trip.   The cost of forwarding let-
ters from Circle City to Dawson City is one dollar
for each letter and two for each paper, the mails
being sent over once a month.   The Chilkoot Pass
is crossed with the mail by means of Indian carriers.   On the previous trips the carriers, after fin-
ishing the pass, built their boats, but they now have
their own to pass the lakes and the Lewes River.

In the winter transportation is carried on by means
of dogsleds, and it is hoped that under the present
contracts there will be no stoppage, no matter how
low the temperature may go.   The contractor has
reported that he was sending a boat, in sections,
by way of St. Michael, up the Yukon River, to be
used on the waterway of the route, and it is thought
much time will be saved by this, as formerly it
was necessary for the carriers to stop and build
boats or rafts to pass the lakes.

Contracts have been made with two steamboat
companies for two trips from Seattle to St. Michael.

VIEW OF WRANGELL (FROM CHIEF'S HOUSE).

When the steamers reach St. Michael, the mail will be transferred to the flat-bottomed boats running up the Yukon as far as Circle City. It is believed the boats now run further up.

The contracts for the overland route call for only first-class matter, whereas the steamers in summer carry everything, up to five tons, each trip.

Sledges and Dogs.—The sleds are heavy and shod with bone sawed from the upper edge of the jaw of the bowright whale. The rest of the sled is of spruce and will carry from six to eight hundred pounds. The sleds used in the interior are lighter and differently constructed. They consist of a narrow box four feet long, the front half being covered or boxed in, mounted on a floor eight feet long resting on runners. In this box the passenger sits, wrapped in rabbit skins so that he can hardly move, his head and shoulders only projecting. In front and behind and on top of the box is placed all the luggage, covered with canvas and securely lashed, to withstand all the jolting and possible upsets, and our snow shoes within easy reach.

An important item is the dog-whip, terrible to the dog if used by a skillful hand and terrible to the user if he be a novice; for he is sure to half strangle himself or to hurt his own face with the business end of the lash. The whip I measured had a handle nine inches long and lash thirty feet, and weighed four pounds. The lash was of folded and plaited seal hide, and for five feet from the handle measured five inches round, then for fourteen feet it gradually tapered off, ending in a single thong half an inch thick and eleven feet long. Wonderful the dexterity with which a driver can pick out a dog and almost a spot on a dog with this lash. The lash must be trailing at full length behind, when a jerk and turn of the wrist causes it to fly forward, the thick part first, and the tapering end continuing the motion till it is at full length in front, and the lash making the fur fly from the victim. But often it is made to crack over the heads of the dogs as a warning.

The eleven dogs were harnessed to the front of the sled, each by a separate thong of seal hide, all of different lengths, fastened to a light canvas harness. The nearest dog was about fifteen feet from the sled, and the leader, with bells on her, about fifty feet, the thongs thus increasing in length by about three feet. When the going is good the dogs spread out like the fingers of a hand, but when the snow is deep they fall into each other's tracks in almost single file. As they continually cross and recross each other, the thongs get gradually plaited almost up to the rearmost dog, when a halt is called,

A TEAM OF DOGS AND DOG SLEDGES.

the dogs are made to lie down, and the driver care-
fully disentangles them, taking care that no dog
gets away meanwhile.   They are guided by the
voice,. using "husky," that is, Eskimo words:
"Owk," go to the right; "arrah," to the left, and
"holt," straight on.   But often one of the men must
run ahead on snowshoes for the dogs to follow him.

The dogs are of all colors, somewhat the height
of the Newfoundland, but with shorter legs.   The
usual number is from five to seven, according to the
load.

List of prices that have been current in Dawson
City during 1897:

Flour, per 100 lbs	$12.00 to $120.00	
Moose ham, per lb	1.00 to	2.00
Caribou meat, lb	.65	
Beans, per lb	.10	
Rice, per lb	.25 to	.75
Sugar, per lb	.25	
Bacon, per lb	.40 to	.80
Butter, per roll	1.50 to	2.50
Eggs, per doz	1.50 to	3.00
Better eggs, doz	2.00	
Salmon, each	1.00 to	1.50
Potatoes, per lb	.25	
Turnips, per lb	.15	
Tea, per lb	1.00 to	3.00
Coffee, per lb	.50 to	2.25
Dried fruits, per lb	.35	

Canned fruits	.50 to	2.25
Lemons, each	.20 to	.25
Oranges, each	.50	
Tobacco, per lb.	1.50 to	2.00
Liquors, per drink	.50	
Shovels	2.50 to	18.00
Picks	5.00 to	7.00
Coal oil, per gal.	1.00 to	2.50
Overalls	1.50	
Underwear, per suit	5.00 to	7.50
Shoes	5.00 to	8.00
Rubber boots	15.00 to	18.00

Based on supply and demand the above quoted prices may vary several hundred per cent. on some articles at any time.

Fare to Seattle by way of Northern Pacific, $81.50.

Fee for Pullman sleeper, $20.50.

Fee for tourist sleeper, run only west of St. Paul, $55.

Meals served in dining car for entire trip, $16.

Meals are served at stations along the route a la carte.

Distance from New York to Seattle, 3,290 miles.

Days required to make the journey, about six.

Fare for steamer from Seattle to Juneau, including cabin and meals, $35.

Days, Seattle to Juneau, about five.

Number of miles from Seattle to Juneau, 725.

Cost of living in Juneau, about $3 per day.

Distance on Lynn Canal to Healey's Store, steamboat, seventy-five miles.

Number of days, New York to Healey's Store, twelve.

Cost of complete outfit for overland journey, about $150.

Cost of provisions for one year, about $200.

Cost of dogs, sled and outfit, about $150.

Steamer leaves Seattle once a week.

Best time to start is early in the Spring.

Total cost of trip, New York to Klondike, about $667.

Number of days required for journey, New York to Klondike, thirty-six to forty.

Total distance, New York to the mines at Klondike, 4,650 miles.

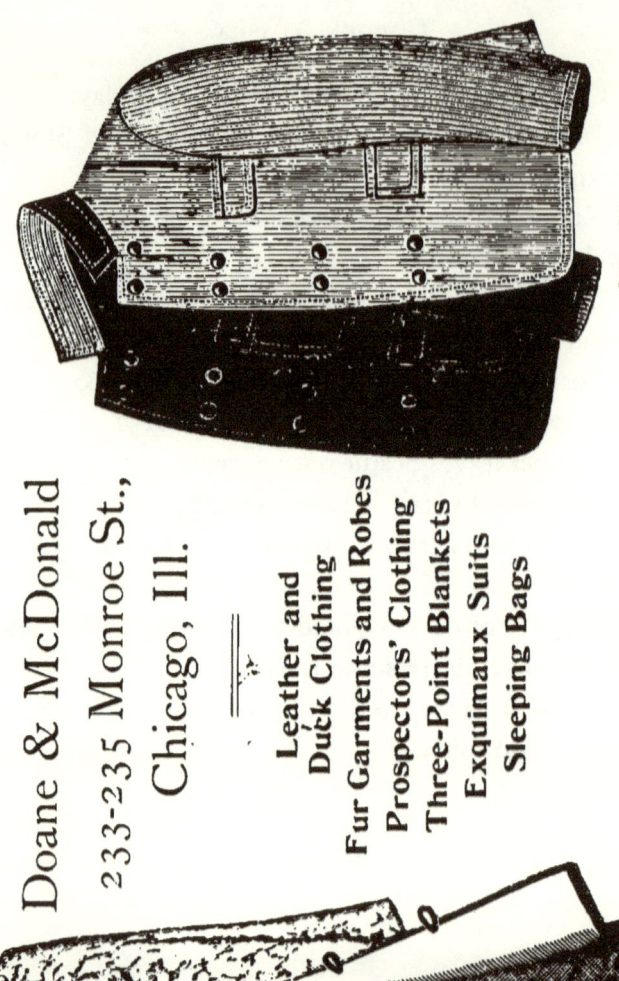

NO. 21.

Doane & McDonald

233-235 Monroe St., Chicago, Ill.

Leather and Duck Clothing

Fur Garments and Robes
Prospectors' Clothing
Three-Point Blankets
Exquimaux Suits
Sleeping Bags

No. 477.

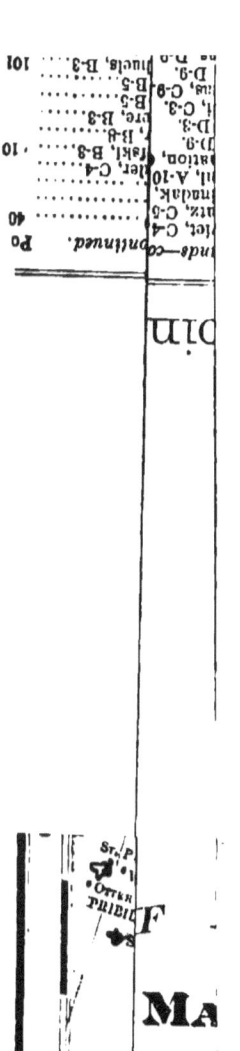

* Money Order Offices.   ¶ Post Offices

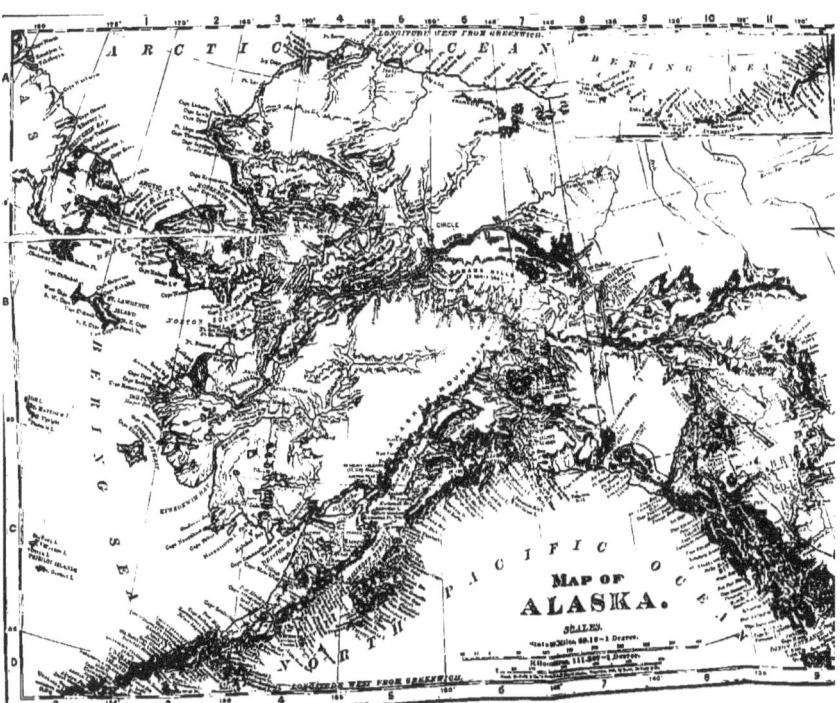

ALASKA.

Districts, Capes and Points, Islands, Lakes, Mountains, Rivers, and Towns.

Towns.   Pop.

Addenda.   Pop.

For Convenient Reference.

MARAH ELLIS RYAN'S WORKS.

A FLOWER OF FRANCE.

A Story of Old Louisiana.

The story is well told.—*Herald, New York.*
A real romance—just the kind of romance one delights in.—*Times, Boston*
Full of stirring incident and picturesque description.—*Press, Philadelphia.*
The interest holds the reader until the closing page.—*Inter Ocean, Chicago*
Told with great fascination and brightness. * * * The general impression
delightful. * * * Many thrilling scenes.—*Herald, Chicago.*
A thrilling story of passion and action.—*Commercial, Memphis.*

A PAGAN OF THE ALLEGHANIES.

A genuine art work.—*Chicago Tribune.*
A remarkable book, original and dramatic in conception, and pure and
noble in tone.—*Boston Literary World.*
REV. DAVID SWING said:—The books of Marah Ellis Ryan give great
pleasure to all the best class of readers   "A Pagan of the Alleghanies" is
one of her best works; but all she writes is high and pure. Her words are all
true to nature, and, with her, nature is a great theme.
ROBERT G. INGERSOLL says:—Your description of scenery and seasons
—of the capture of the mountains by spring—of tree and fern, of laurel,
cloud and mist, and the woods of the forest, are true, poetic, and beautiful.
To say the least, the pagan saw and appreciated many of the difficulties and
contradictions that grow out of and belong to creeds. He saw how hard it is
to harmonize what we see and know with the idea that over all is infinite
power and goodness * * * the divine spark called Genius is in your brain.

SQUAW ÉLOUISE.

Vigorous, natural, entertaining.—*Boston Times.*
A notable performance.—*Chicago Tribune.*
A very strong story, indeed.—*Chicago Times.*

TOLD IN THE HILLS.

A book that is more than clever. It is healthy, brave, and inspiring.—*St
Louis Post-Dispatch.*
The character of Stuart is one of the finest which has been drawn by an
American woman in many a day, and it is depicted with an appreciation
hardly to be expected even from a man.—*Boston Herald.*

IN LOVE'S DOMAINS.

There are imagination and poetical expressions in the stories, and readers
will find them interesting.—*New York Sun.*
The longest story, "Galeed," is a strong, nervous story, covering a wide
range, and dealing in a masterly way with some intricate questions of what
might be termed amatory psychology.—*San Francisco Chronicle.*

MERZE; The Story of an Actress.

We can not doubt that the author is one of the best living orators of her
sex. The book will possess a strong attraction for women —*Chicago Herald.*
This is the story of the life of an actress, told in the graphic style of Mrs
Ryan. It is very interesting.—*New Orleans Picayune.*

Alaska-Klondike
Gold Mining Company

CAPITAL STOCK...500,000 Shares.

Par Value...$10.00 each.

Full Paid—Non-Assessable.

This Company is a
Transportation,
Commercial, and Mining Corporation
owning large GOLD GRAVEL claims on the Yukon,
Klondike, and other rivers in Alaska, and now have
under construction steamers to ply on the Yukon
next season.

The Board of Directors are a sufficient guarantee that the
affairs of the Company will be well managed.

DIRECTORS.

JAMES RICE,
Late Secretary State of Colorado.

WM. SHAW,
Capitalist, Chicago.

E. M. TITCOMB, Vice-Pres't and Gen'l Manager,
Eastman Fruit Despatch Co.

H. C. FASH,
Member Maritime Exchange, New York.

GEO. W. MORGAN,
Circle City, Alaska.

A limited amount of Shares are offered at **$10.00 per share.**
For information, address,

Alaska-Klondike Gold Mining Co.

96 BROADWAY, NEW YORK.

HON. JAMES RICE, PRESIDENT.
W. L. BOYD, SECRETARY.

www.ingramcontent.com/pod-product-compliance
Lightning Source LLC
Chambersburg PA
CBHW020011030726
47500CB00002B/532